So You're a Doctor Now

CHARLES HOWARD

Copyright © 2013 Charles Howard.

All rights reserved. No part of this book may be reproduced, stored, or transmitted by any means—whether auditory, graphic, mechanical, or electronic—without written permission of both publisher and author, except in the case of brief excerpts used in critical articles and reviews. Unauthorized reproduction of any part of this work is illegal and is punishable by law.

ISBN: 978-1-4834-0288-8 (sc)
ISBN: 978-1-4834-0290-1 (hc)
ISBN: 978-1-4834-0289-5 (e)

Library of Congress Control Number: 2013913432

Because of the dynamic nature of the Internet, any web addresses or links contained in this book may have changed since publication and may no longer be valid. The views expressed in this work are solely those of the author and do not necessarily reflect the views of the publisher, and the publisher hereby disclaims any responsibility for them.

Lulu Publishing Services rev. date: 9/09/2013

TABLE OF CONTENTS

The Oath of Maimonides ix

Note for published edition xi

Introduction 1

Two Pillars 7
    CARBON FIBRE PLATES 11
    WHY BECOME A DOCTOR? 15
    WHAT IS A "GOOD DOCTOR"? 16

The diagnosis 22
    APPENDICITIS? 22
    PREVENTION IS BETTER THAN CURE 23
    TRAUMA CLINIC 24
    HAEMORRHOIDS 26

Be careful what you say 28
    WHY MEN DON'T MAKE GOOD "AGONY AUNTS" 28
    BEDRIDDEN 29
    DOES HE TAKE SUGAR? 31
    MOTHER LOVE 32
    ONLY A MOTHER 32
    THESE THINGS HAPPEN 33
    WEIGHT LOSS 34
    PEOPLE ARE EASILY INSULTED 35
    ASHKENAZI/SEPHARDI 36
    SELF-ESTEEM 37
    BELITTLING PEOPLE 39

    ASKING QUESTIONS 40
    HISTORY 42
    TELLING LIES 46
    MY WORD IS MY BOND 48
    PARDON? 50
    DAVENING (PRAYING) 50
    HEAD INJURY 51
    TOMATOES 52
    ARCHIE 53
    OLD DAUGHTER 55
    THE OTHER SIDE OF THE TABLE 56
    LOSS OF CARING IN OUR PROFESSION 57

Patients are optimistic **59**
    SHE WILL GET BETTER, WON'T SHE? 59
    "'OOVERING" 60
    SPINAL INJURY 61

Grateful Patients **64**
    A TIN OF BISCUITS 64
    BACK PAIN 66

Our worst nightmare **70**
    ELBOW FRACTURE 70
    SUBLUXING PATELLA 72
    SOMETIMES YOU JUST NEED TO BE LUCKY 76
    AND SOMETIMES YOU HAVE JUST GOT TO EXAMINE THE
    PATIENT CAREFULLY! 78

The mind and physical illness **81**

Research **87**
    THE ABC TRAVELLING FELLOWS 87
    PREPARE YOUR PROJECT 88
    DIFFERENT PEOPLE INTERPRET THE SAME THING
    DIFFERENTLY. 89
    DIGITAL ARTERY ANEURYSM 92

    METEORITES 93
    FAMILY 94

It's not interesting 99
    ADVICE NO 1 100
    ADVICE NO 2 100
    ADVICE NO 3 102
    ADVICE NO 4 102
    THE CHAIR 103

THE OATH OF MAIMONIDES

"The eternal providence has appointed me to watch over the life and health of Your creatures. May the love for my art actuate me at all time; may neither avarice nor miserliness, nor thirst for glory or for a great reputation engage my mind; for the enemies of truth and philanthropy could easily deceive me and make me forgetful of my lofty aim of doing good to Your children.

May I never see in the patient anything but a fellow creature in pain.

Grant me the strength, time and opportunity always to correct what I have acquired, always to extend its domain; for knowledge is immense and the spirit of man can extend indefinitely to enrich itself daily with new requirements.

Today he can discover his errors of yesterday and tomorrow he can obtain a new light on what he thinks himself sure of today. Oh, God, You have appointed me to watch over the life and death of Your creatures; here am I ready for my vocation and now I turn unto my calling"

Attributed to, but probably not by, Maimonides

Maimonides[1]

---

[1] http://de.wikipedia.org/wiki/Datei:Maimonides-2.jpg
Attrib. Blaisio Ugolino (1744) Thesaurus antiquitatum

NOTE FOR PUBLISHED EDITION

This book was originally written for my son, who has recently qualified as a doctor. He, and some of his friends who read it, felt it may be of interest to a wider audience and so he and I decided to publish it. Although in the original, the real names of the people and places involved were used, for obvious reasons, we have hidden their identity by the use of pseudonyms.

For those readers who are unfamiliar with the British system of medicine, surgeons bear the title of Mr. rather than Dr. and I have used this when referring to surgeons. The title Mr. dates back to the middle ages when surgery (consisting mainly of lancing abscesses, treating haemorrhoids, battle field wounds, amputations, dental extractions, blood lettings), was completely separate from medical diseases. No self-respecting physician of the time would consider sullying his hands with surgical procedures and these were originally carried out by barbers. There gradually emerged a profession of Barber-Surgeons, and it was only about two hundred and seventy years ago (1745) that this guild split into Barbers and Surgeons. However to this day, once admitted to the Royal College of Surgeons, the young doctor gives up his title of doctor, to become Mister. Surgeons are still held by physicians, at least in fun, to be a lower intellectual entity; as expressed by the following:-

Physician's definition of an operation:-

"The half-asleep watching the half-awake being half-murdered by the half-witted."

Physician: "What are the qualifications to be an orthopaedic surgeon?"

Answer: "Strong as a horse and twice as bright, but since the advent of power tools they don't actually have to be so strong."

As most of the stories occurred in Britain, the English style spelling has been retained for consistency, although the American style is often the most logical.

INTRODUCTION

This book is for you, Gadi, from me. I dedicate it and give it to you in its entirety. This is not because I love your brother and sisters less, but partly because you have decided on a medical career and many of the incidents will be relevant to your dealings with your patients, and partly because, as the first born, you suffered from our mistakes more than the others; but more of this at the end of the book.

Anyway, I have decided to write for you some of the incidents/stories, that have happened to me as I traversed through my medical career (and outside of it). Some were told to me and some actually happened to me. These are the ones I remember. I am sorry now I didn't keep a diary or at least recorded other incidents which I am sure happened but I have now forgotten.

Some I may have already told you, nevertheless I think it is worth putting them down in writing.

But to start off with, I want you to know how proud both Mum and I are of you. For me becoming a doctor was rather easy, I got accepted straight into University after school (even though I didn't get good enough grades first time round, they held the place open for the next year for me to improve my exam results). You had to go through five years army service, a Biology degree and then a Master's degree, before even starting the long and arduous path up the medical ladder. Before you is the real school; life with real people, real diseases, real signs and symptoms. You will find it difficult, exhilarating, tiring, irritating, worrying, frustrating and intensely satisfying. You will feel the thrill of the intimate relationship between yourself and your patients, and the frustrations when things don't go right.

So now the third generation of our family takes on the mantle, the responsibility of looking after our fellow human beings.

Alex Howard F.R.C.S. 1911-1996. (Grandfather) General surgeon and family practitioner (G.P.), aged 68

Charles Howard F.R.C.S 1950—Orthopaedic surgeon aged 63

Gadi Howard M.D. 1977- Just qualified, aged 36

Being a doctor is an honour and privilege, and often abused and unappreciated by so many of our colleagues. You will see much arrogance and much self-opinionated behaviour from them. This is well illustrated in a comment on an online article about medical training that appeared in the Daily Telegraph, which included the phrases "abysmal quality of many of the doctors" and "their standards are so low, their indifference and clinical incompetence so casually evident and they obviously couldn't give a sh*t about their patients."

I can certainly understand the obvious frustration and contempt with which this commentator holds for doctors. As you may know I produced a film for our junior doctors, "The art and practice of Plaster fixation", partly because they did this so badly and I was constantly meeting children a day or so after they had been treated in awful plasters. But what eventually decided me that I had to make this film, was when I saw the child (depicted in picture below) on my daily ward round, the day after he had had his fracture reduced and plastered and admitted overnight for observation.

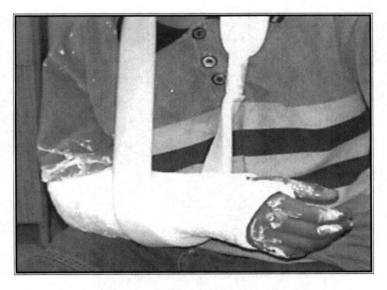

The state in which a young boy had been left in by the doctor who had treated his fracture and the nursing staff who had looked after him overnight.

The child was fine, the reduction of the fracture excellent and the plaster well applied, but as you can see he had been LEFT OVERNIGHT with his fingers and clothes covered in plaster. The disrespect for the patient; no let's not call him "patient" but rather a fellow human being, a child who had come to be treated by these doctors and nurses, is a disgrace to the doctor and nurses, the hospital, the Orthopaedic specialty and the medical profession. I cannot tell you how disgusted I was and even now when I think about it, it makes me absolutely furious. This is exactly what was being referred to in the Daily Telegraph comment. Perhaps if it had been an isolated event, I could have just calmed down and forgotten about it, but I guarantee you if you watch patients being discharged from any casualty in the vast majority of hospitals throughout the world, you will see this repeating itself over and over again. Doctors are taught to treat fractures excellently but for some reason they seem to forget these fractures are part of a person. How long does it take just to wipe the fingers or toes clean? 20-30 seconds? Certainly pressure of work is not responsible nor does it excuse leaving people in a plaster soaked state (which is uncomfortable as well as unaesthetic).

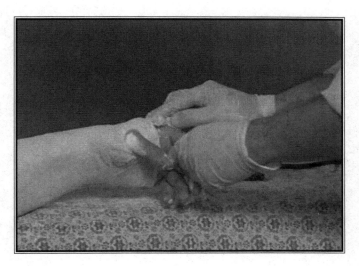

After finishing applying the plaster, take a wet tissue/cloth and clean up the patient from any excess plaster, dirt, blood that may still be on his arm[2].

---

[2] from "The Art and Practice of Plaster Casting" by the author

It makes you wonder sometimes why our profession is held in such esteem by the population. And, in spite of the comment quoted above, make no mistake about it, we are highly valued. What we say and how we say it have enormous effect on our patients; but more of that later.

TWO PILLARS

Where to start? Well, first welcome to the profession; and a very old profession it is. In spite of the claims by the Ladies of the night that theirs is the oldest profession, medicine far and away predates them. Even before Homo sapiens appeared, the early hominids, and possibly even earlier species, were doing medical things to each other in an attempt to cure the numerous afflictions from which they suffered.

In my opinion Medicine stands on the two pillars of Humanity and Science, i.e. the doctor patient relationship and the Science and technology that has produced the drugs, imaging techniques, instruments and operating procedures that doctors employ to treat their patients.

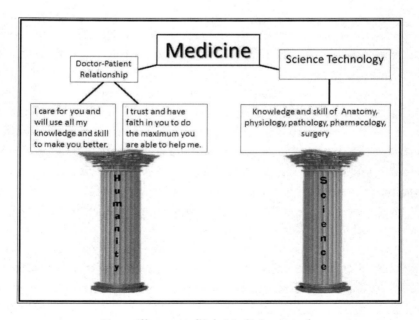

Two pillars on which Medicine stands

On the left there is a pillar which represents the doctor-patient relationship, which hasn't changed since the practice of Medicine started. The right hand pillar represents the tools used by medical personnel to treat their patients. With advancing science, this is changing all the time.

Until about one hundred and fifty years ago the Science pillar was virtually nonexistent and therapies consisted of ineffective herbs, poultices, enemas and a variety of procedures some of which were positively dangerous, e.g. bloodletting (of which I will say more later in the book) and trephining. Even recently, strange modalities, such snail mucus, ground up mummy powder, and tobacco smoke enemas were used.

But before you laugh too loudly at these strange, ignorant therapies used by our predecessors, I should point out that there are still plenty of such equally odd treatments around today. Wikipedia[3] gives over one hundred and fifty examples of such treatments each of which have many practitioners and followers; a few examples include:- acupressure, prayer, bee stings, astrology, flower therapy, bioresonance therapy, spa bathing, use of coloured lights, magnetic therapy and drinking human urine!

So, why, in this modern day and age, when excellent scientific drugs and surgeries are widely available, often provided for free by the state, do so many people turn to, and pay for these alternate methods?

One would have thought fewer and fewer people would subject themselves to treatments such as drinking their (and other people's) toxic waste, staring at the sun, foot massages, homeopathy or coffee enemas. A Google internet search of "Alternative medicine" produced about forty five million results, showing these are thriving and more and more people are turning to them. Why? What is the attraction? What are we doctors doing to drive people to these, at best benign, at worst positively dangerous, activities?

Numerous surveys and articles have shown people are unhappy with the lack of time and consequent communication with their doctors. Going to an alternative practitioner almost always involves payment

---

[3] http://en.wikipedia.org/wiki/List_of_branches_of_alternative_medicine

and full attention on the *person* (no blood tests, x-rays, ultrasounds, CT, MRI, urine analysis, faecal examination, ECHO, stress tests etc. to take up the available time for these practitioners). Patients then get actual, on the spot, hands on, instant treatment whereas a doctor will give a piece of paper for some chemical to be taken three times a day at home or a referral to a physical therapist for treatment at a later date, again probably after a long wait for an appointment. It seems as if patients feel they have lost the first pillar and the second pillar involves some strange and incomprehensible procedures which their doctor can't, doesn't, or won't explain.

According to one such survey, "Patient-Doctor Global Communication Assessment" of over twenty two thousand patients in twenty three countries[4] between 96% and 50% of patients were dissatisfied with their communication with their doctor. Only a third thought their doctor treated them with respect, spent enough time with them, showed they cared, involved the patient in treatment decision-making (doctor knows best!), and didn't use medical jargon that the patient simply couldn't understand.

Horrendous, humbling figures!

As I hope you will understand after reading this book, it is we doctors that are responsible for the poor results shown in the above survey. The first pillar which I have called Humanity is often lacking or completely missing from the interaction with our patients. This is partly from lack of time, partly because we have to concentrate so hard and be fully focused on the technical side, that the patient, the cause and most important part of the relationship, is often forgotten. Many patients complain that when going in for a consultation the doctor doesn't even look at them, he is too busy typing on the computer, looking up blood results, examining x-rays etc. And yet, over 4 million years of medical practice succeeded, without the second pillar, based simply on eye contact, verbal communication and then physical contact. Firstly the physician creates a bond between himself and his patient by looking into his eyes, recognizing his existence. He deepens it by talking to and

---

[4] WIN Worldwide Independent Network of Market Research 2010 http://www.trig-us.com/images/stories/spotlight/gpdcomreport.pdf

about his patient, by touching, probing and exploring his patient's body, illness, and problem. He absorbs the patient's psyche, reacts and cares for this person who has sought his help. Add in the Placebo Effect[5], plus the fact that most minor conditions get better by themselves and you can understand how support and caring alone is so sort after and why the alternative practitioners are so popular.

I have to say in our defense, it isn't wholly our fault. The number of patients waiting to be seen often means we have very limited time with each patient. But it takes time, back and forth communication, to get to know each other, for the patient to know he can trust his doctor, for the doctor to understand what's going on with his patient. That being said, however, it takes just as much time to smile welcomingly to the patient as it does to scowl irritably at the ceiling. It takes just as much time to say "That's alright, don't worry about it" as it does to say "Well next time, make sure you *do* get here on time" when the patient is a few minutes late for his appointment. It doesn't take more time to show you care than it does to show you don't. It just takes more effort.

The various diagnostic procedures will also take a big bite out of the precious, limited time you have with each patient. Once you have greeted the patient, gotten a good history, carried out a thorough examination, and decided on and ordered a variety of diagnostic tests, eventually you will come to a diagnosis and will want to start treatment. This you now have to explain to the patient, particularly difficult if there are several options. Most patients don't have a good grounding in anatomy, physiology, pathology, pharmacology, etc., which forms the basis of and is the rationale for treatment, making it difficult for them to be involved with the decision-making. And according to the survey mentioned above, being involved in decision-making is very important to them.

How you pitch the options will grossly affect their decision as the next story indicates.

---

[5] In clinical trials, an active drug/treatment is tested against an inactive one (e.g. the same looking tablet filled with chalk or sugar). 30% of patients on the inactive regime will report their symptoms have resolved or greatly improved. The mere fact of having "treatment" works!

Carbon fibre plates

We were carrying out a trial of a new concept in fracture treatment, Carbon fibre reinforced plastic plates (CFRP).

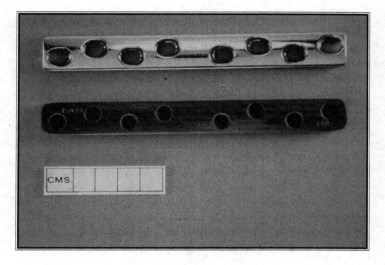

On the top is a stainless steel plate and below is a CFRP plate

But first a bit of background as to why we were trying out these plates. Many tibial and forearm fractures are problematic. Plaster treatment of tibial fractures may cause deformity, shortening, loss of muscle, and stiff painful joints. When first introduced, metal plates enabled accurate alignment, but rapidly broke due to metal fatigue secondary to small movements between the fractured fragments. In 1948, by perfect reduction and compressing the fragments together and plating, these movements were inhibited and the plates didn't break. A whole industry arose in Switzerland (the AO group). They produced beautiful instruments and plates and screws, and developed surgical techniques using this philosophy. It swept the orthopaedic world. Special courses were held to teach the method and all over the world orthopaedic surgeons took on this method and it was considered almost blasphemous even to criticize it. However, with all its advantages, there were numerous disadvantages; it was technically difficult as absolute perfect reduction of the fracture was essential which required wide dissection of the bone, much stripping of muscles and destruction of the bone's blood supply. By the end of this operation it had to appear as if there was no fracture and the plate applied with such strong compression that absolutely no movement between the fragments could take place. But, it is the ***movements*** between the fragments that tells the body to heal the fracture by producing a ball of callus (new bone) around the fracture. Preventing these movements inhibits the formation of this callus and can delay or prevent the fracture from healing.

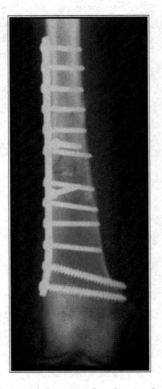

In this x-ray, a femoral fracture has been plated with an AO plate. The reduction is perfect but there still is a small gap between the fragments as seen by the slightly darker lines between the fragments. This allowed small movements to occur.

So You're a Doctor Now

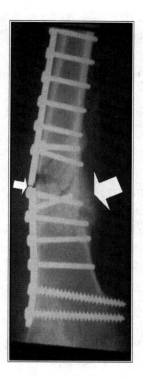

After a few weeks of continual small movements at the fracture site the plate has broken from metal fatigue (thin white arrow). However, some new bone (thick white arrow) has been produced but insufficient to support the fracture.

To allow small movements at the fracture site, without plate breakage, carbon fibre reinforced plates were developed, tested in sheep and a clinical trial in human fractures undertaken. It was decided that this should not be a randomized trial but patients should have the right to choose which method they wanted. You would imagine from reading the above information that most patients would choose the CFRP plate, and so they did until the arrival of a new house officer (intern) whose job it was to explain the choices and get consent. After she arrived all the patients chose the stainless steel plate. For some unknown reason she had taken a dislike either to the consultant running the trial, or the plate itself, I don't know which, but after explaining why the patient's fracture needed surgery, this is how she offered the alternatives to the patients.

"We are trying out a new plate that we are hoping will give better results, but we really don't know yet as it has only been used a few times and we are still learning the problems. The carbon fibre plate allows the fracture to wobble, which is said to improve fracture healing. Apparently, the pain from this movement stops, or at least is much reduced within a week or so. The movements of the fracture does cause some rubbing and chaffing between plate and screws, leading to the release of small black particles of carbon and plastic in the surrounding tissues. Although it does cause some black staining we have carried out trials in rats to show this material doesn't cause cancer and although humans are different from rats we believe it is safe in this regard. The alternative is a stainless steel inert plate that has been used by virtually every trauma centre in the world for over thirty years. It is claimed that it can slow healing, although I personally haven't seen any problems from it. Which would you prefer?"

But doctors are also human. After years of studying, working hard to become, for example, an orthopaedic surgeon, a doctor's interest and skill is the arthritic knee, the fractured forearm, the painful shoulder, etc. He has neither the time, training nor desire to get involved with any ear disease, social problems, depression or psychological make-up of the patient. Furthermore, while the patient wants to be respected as a human being he/she also wants his arthritic knee, fractured forearm, painful wrist or whatever to take the full attention of the orthopaedic surgeon and doesn't expect or want him to be dealing with his hypertension, irritable colon or eczema. Later in the book I have related a story about a depressed female patient on an orthopaedic ward, how the Consultant Surgeon treated her and how I think he should have handled her.

BUT no matter how many patients are still waiting, no matter how hard pressed you are with a difficult clinical problem, there is time, there is ALWAYS time, to show you care and "use" the first Pillar.

Regarding the Science pillar, as shown below, it is humbling to note that very few of the medical advances were made by doctors.

1796	Edward Jenner starts Vaccination	Physician
1804	Fredrick Serturner discovers Morphine	Pharmacist
1820	Labarraque started disinfection by chlorination	Chemist
1865	Lister introduces asepsis in surgery	Surgeon
1870	Pasteur germ theory of disease	Chemist, microbiologist
1895	Roentgen discovers X-rays	Physicist
1880-1935	Urban sewage systems built	Engineers/politicians
1928	Fleming discovers penicillin	Physician
1951	Carr invents MRI	Physicist
1953	Watson, Crick, Franklin Gosling discover DNA structure	Microbiologist, biophysicist
1966	Koa and Hockham produce high grade optic fibres	Electrical engineers
1984	Human Genome project start	Molecular biologists

A list of some of the important medical advances, their discoverers and occupations. Of all these it was the introduction of a proper sewage system that has saved more lives than all other medical technologies together.

Why become a doctor?

Why do people become doctors? I don't know what propelled you to this course. For me, it originated in my childhood.

When I was nine or ten years old, I was in prep. school. In my class there was a boy called Jonnie. Jonnie was bigger than the rest of the class, ginger haired, wore glasses, had a bit of a speech impediment and was very clumsy. In retrospect, I think he probably had a mild form of ataxic cerebral palsy. He had a lovely personality and was well liked by the whole class. However when it came to choosing teams during the play breaks for "Cowboys and Indians" or "Cavaliers and Roundheads" (games which as all schoolboys know consist of galloping at the opposite team on pretend horses shouting "Charge" and waving imaginary swords or shooting imaginary pistols-"phe-aw, phe-aw" at each other), Jonnie was never chosen. I always went with him and a kid called Robert who was, to use today's PC expression, intellectually challenged. This was not because I liked them more than the rest of the class did; as I said before we all liked them, but because I knew it was my job to look after Jonnie. I don't know why or how I knew this, I just did.

I am sure there are many reasons behind the decision to become a doctor, different for each person. BUT, if you ask me, I think the reason is simple; to look after people.

So, I wish you every success in your career and, whatever it means;

Be a Good Doctor.

What is a "Good doctor"?

While I was in Cardiff I had a discussion with a friend and colleague, Ian, about what we thought made a good surgeon. We were both orthopaedic registrars at the time. We, of course, were by no means the first to have this discussion and although much has been talked and written on the subject over the ages, there still is no precise answer. Anyway, we came up with five criteria.

1) Caring. That might sound obvious and trite, but you will find it is very easy for patients to become routine cases, particularly the minor common conditions which, after a while, are no longer so interesting. But each individual patient must remain, exactly that. It is the PATIENT that is the important aspect not the boring (or interesting) complaint he has come with. It has been said the caring professions should not become emotionally involved with their patients. Nothing could be further from the truth. If you are not emotionally involved with your patients, become a bank clerk.

So what do patients want? Obviously they want to be made better; he/she wants the correct diagnosis and treatment. But there is much more to it than this. Some patients may want information about their condition. If you can understand what's happening, it is far less frightening. Others prefer to remain completely ignorant and pretend it doesn't exist (head in the sand) and of course there are those who lie somewhere in between. You have to recognize which sort of person you are dealing with; not always easy. They also need you as a friend who really cares about and is interested in them.

I use the word 'caring' in the widest sense. If you have a genuine concern for your patients you will care enough to learn and know as much scientific medical knowledge as possible, you will care enough to get a good history and examine the patient properly. It is this caring reflex you must develop. I think at the start of our medical careers all young doctors have it to some extent or other, but over time the pressures of work, the routine nature of much of it, (as illustrated by the comment in the Daily Telegraph at the beginning of this monograph), can crush it and leave many doctors void of appearing to care. Never let this happen to you. Nurture the reflex like a delicate plant- nurture it daily, let it grow and never let it wither.

2) Knowledgeable about medical illnesses and surgical techniques and up-to-date with new (and old) advances. You may think this is obvious. But there are, as I am sure you know, so many possible diseases, syndromes, conditions it is hard, even impossible, to know all about all of them and keep this information constantly in your head. But, if you haven't learnt about them and don't keep up to date with any new knowledge in the first place, you definitely won't be able to access the information when you have a patient in front of you and you want to make a diagnosis or select the correct treatment. It is too easy to cut corners, particularly when you are experienced and feel you can make the diagnosis from just a cursory history and quick examination.

Just a tiny sample of medical books.

3) A good clinician. By this, we meant the ability to extract correct and relevant information from the patient and carry out a proper examination and the ability, much like a good detective, not just to look but to actually *see* the physical signs present. As you will find out, this is not easy. Unfortunately, physical signs, particularly early on in the disease, may be very minimal and subtle. Is there a slight thickening of the synovium of the patient's knee or is it just the natural asymmetry between the two legs? I have recounted a couple of examples of this later on. By the way, Conan Doyle based Sherlock Holmes on a friend, a doctor, who had this wonderful ability to OBSERVE.

So You're a Doctor Now

The Sherlock Holmes stories first appeared in the Strand magazine and were illustrated by Sidney Paget. This one, from "The adventure of the Norwood builder" shows Holmes examining a bloodied fingerprint.

"I knew it had not been there the day before. I pay a good deal of attention to matters of detail, as you may have observed. Therefore, it had been put on during the night."

4) A good technician (we were discussing Surgery). It is important to be fast, accurate, and gentle but decisive in handling tissues in surgery. For this, an intimate 3D knowledge of human anatomy is essential. But it is not a race or a competition with other surgeons. Unfortunately there is a tendency for surgeons to judge other surgeons, not on their results, which are very difficult to know and assess, but on "how fast they are." Later in this monograph, I have written about two surgeons; Mr. Cook and Mr. Simmons. The two were the complete opposite.

Mr. Cook was an incredibly slow surgeon and "messed about", having difficulty deciding whether to cut a piece of tissue; was it a small nerve or blood vessel or just a piece of connective tissue? I remember once assisting him in a hemi-colectomy which took 10 hours (although normally it took him only six hours!). Mr. Simmons was the fastest, most decisive surgeon I have ever come across. He would routinely do a total hip replacement in 40-45 minutes. He had a superb eye and was extremely accurate most of the time. I say most of the time because of his personality (more on this later) and over confidence. He once (in a private patient) filled the femoral canal with cement and inserted the femoral prosthesis, reduced the hip before the cement set (a reasonable technique as the pressure from the reduced hip keeps the prosthesis in the natural position as the cement hardens and expands) and closed the wound while the cement was setting. It was only later, on looking at the post-operative x-rays, that it became apparent the prosthesis was <u>outside</u> the femoral shaft. Mr. Cook was a Sherlock Holmes, clinically. Mr. Simmons, on the other hand....Well, more of that later.

5) A good researcher. We added this in because if you do research, you are constantly thinking, assessing, and asking, is it right, maybe it could be better, are the results really as good as I think?

What we (Ian and I) agreed was that nobody is going to get 10 out of 10 in all five groups. We reckoned each doctor should appraise himself according to our list and go where his strengths lie.

Even so, it is difficult to know what a good doctor is, as the following narrative shows. When I was in Lewisham [I was there for one year doing six months surgical house job (internship) and six months as a casualty officer], I came across one of the local general practitioners, a doctor by the name of Dr. Graham. We (the medical staff) thought he was a superb doctor. Before sending in a patient to casualty he would ring and speak to the casualty officer. Not infrequently he brought the patient in his own car. We used to joke that if Dr. Graham sent in a patient with an appendicitis, you needn't bothering examining the patient, just book theatre because he was inevitably right. His patients, however, weren't so enthusiastic. He was a rather dour chap, rarely smiling. He didn't give antibiotics (even in those days) to everyone with a sore throat, but took a throat swab first. He wouldn't give you an off-work sick note unless you were ACTUALLY sick and so on.

In comparison, there was a GP near the hospital at which I trained, who was extremely well liked by his patients and equally well disliked by all the hospital staff. You wanted an X-ray, he gave you a form; you needed a few days off work, no problem-have a sick note. Need a referral to hospital? Just ask. In fact on one occasion he was woken at 11:30 at night by a patient (he lived in a flat above his surgery). Opening his window he asked what the patient wanted. "Oh, I got this terrible pain in me belly, doc," came the reply. "Hang on a minute," said his doctor and left the window, only to return, as promised, one minute later. He threw down a letter and told his patient to take it to casualty, which he did. The letter read:-

"? Appendicitis. Please see and advise."

Although the casualty officer got into some trouble for his reply, in my opinion, it borders on perfect.

He wrote back, "Have seen, advise you to do the same."

So here are two doctors; one who cares, is an excellent clinician but has lousy empathy with his patients and one who cares, is a lousy clinician but has great empathy with his patients. Which one is the good doctor?

I think the answer has to be, neither of them.

THE DIAGNOSIS

In spite of marvelous advances in imaging and auxiliary tests, diagnosis still depends greatly on proper history and examination.

Appendicitis?

While I was doing my surgical house job (internship), a fourteen year old girl presented to casualty with abdominal pain, vomiting and a mild temperature. She was an extremely pretty young girl well into puberty and like many of her contemporaries followed the then fashion of miniskirts and make-up. On examination she was afebrile and had vague tenderness in the abdomen. She was kept in overnight during which time she improved and was discharged home. A week later she returned with similar symptoms and again minimal findings on examination. Again she was kept in overnight, got better and was discharged, only to return a few days later but this time was indeed pyrexial, toxic and slightly dehydrated from twenty four hours vomiting. She was very tender in the lower abdomen, mostly in the right iliac fossa. She arrived this time a few hours before the consultant's once a week ward round. The consultants did a ward round only once a week to see the post-operative and pre-operative cases; the day to day running of the ward (up to 60 patients) was left to the registrar, possibly a senior registrar, and a house officer. Our consultant was Mr. Cook whom I mentioned earlier. As I stated previously, he wasn't a brilliant technical surgeon but he was a wonderful clinician. We (the registrar and I) had made a diagnosis of appendicitis. When Mr. Cook arrived, he asked what were the findings on PR (Per Rectum- the insertion of a finger into the back passage to examine the lower bowel, prostate/uterus and bladder).

We admitted we hadn't thought it necessary as the diagnosis was obvious and we hadn't wanted to put this young girl through the embarrassment of a PR, either today or on the previous two occasions. He was annoyed with us and insisted on doing a PR on her straight away. This turned out to be very painful for her but he then insisted we all do a PR on her, which I thought at the time cruel and unnecessary. However, on PR I felt, like the registrar, what I thought to be an appendix abscess which was what was causing her so much tenderness on this examination. When we said what we had found, he said, "No, it's about 2cm too medial, call the gynecologists." We were a little cross with him because obviously she had an appendix abscess. Anyway, the gynecologists came and on vaginal examination and needle aspiration, it turned out she had a gonococcal pyosalpinx (an infection in the tube that brings the egg from the ovary to the uterus caused by the Venereal Disease, gonorrhea). In spite her young age she was extremely "active" in the London docks and had contracted the disease from a Greek seaman. When Mr. Cook made us all do a PR on this young girl, it wasn't out of cruelty but to teach us the most valuable medical adage, known to all and ignored by all so many times that stories such as this and the next one, abound in every medical centre in the world :- "If you don't put your finger in, you'll put your foot in it." I have never forgotten it and I am sure, never has the registrar.

Prevention is better than cure

A forty two year old doctor was referred to the surgical outpatient clinic of a large London teaching hospital with a history of abdominal pain, constipation and occasional rectal bleeding. After a minimal physical examination, a Barium enema was ordered. Until this was done two months had passed. It was inconclusive and when he returned to the clinic it was decided to carry out a colonoscopy. He was put on the waiting list and again after a further two/three months wait, a colonoscopy was carried out which revealed a rectal carcinoma. After a further two months wait on the NHS list he was admitted for surgical resection. At surgery it was noticed that the lymph nodes were involved and there was a solitary metastasis in the liver. He was admitted for a hemi hepatectomy (removal of the half of the liver containing the metastasis). In those days liver surgery was just starting and it was a new and huge operation.

The case was presented at one of the surgical teaching sessions, by the consultant who had carried out the resection, emphasizing the various technical problems and difficulties that had arisen during this challenging procedure and how he had overcome them. He concluded, with much self-satisfaction, that there was much to be learnt from this case.

"Yes," piped up a visiting Australian registrar, from the back of the room. "Maybe it will teach you to do a bloody PR when the patient first presents." But the PR is just one example of patients not getting examined properly.

Trauma Clinic

The trauma clinic at the hospital at which I was a registrar in orthopaedic surgery consisted of ten cubicles divided by curtains arranged around three walls of the room in a "U" fashion. Patients were brought into these cubicles with their notes and x-rays and the three or four doctors would go around in a clockwise direction, going into the first cubicle that didn't have a doctor already examining the patient.

The patient would be seen, sent for x-rays, plaster removed or changed, sent to physiotherapy, discharged etc. according to his needs. The patient would be then redirected as instructed, by the nurse and the next patient brought in. It was a very efficient system and up to 160 patients a day could be seen and treated. One morning I went into a cubicle to see a child of about ten or eleven years old. He had been seen the week before in casualty having fallen on his left side and had pain in the left clavicular (collar bone) area. He had been x-rayed and no fracture was seen, reassured and given a follow-up appointment for fracture clinic. I saw the x-rays; indeed no fracture. I felt his clavicle, not swollen tender or deformed. Although it was painful, he moved his shoulder reasonably well and I reassured his mother that there was no break or dislocation and the contusion would get better by itself. The whole consultation had taken less than 3 minutes, which was about average for this very efficient clinic. As I was leaving the cubicle to see the next patient, his mother started to cry and said, "I know there is something wrong with my little boy, and nobody will listen to me." I remember to this day being deeply affected by the way she uttered these words with such intensity and frustration. So I returned and said, "OK, don't cry. Let's have a look at him." At this point we took off his T-shirt. I didn't have to examine him, just take a look. He was very thin, and the supraclavicular fossa, the axilla, inguinal (groin) regions were filled with large, swollen lymph glands. We admitted him straight away and the diagnosis of lymphoma was confirmed. It turned out his cousin had died two weeks earlier from a lymphoma, and in spite of starting treatment, he too died within two weeks.

But at the moment after he removed his shirt and I saw the lymph nodes, I said to myself, from now on I am either going to examine patients properly, not through their clothes, not just via their x-rays, but PROPERLY or I will give up medicine and become a clerk in a bank.

Since then I am happy to say I have not had a similar shocking experience, but nevertheless, one never knows what unexpected pathology is lurking in the next patient to come through the door.

The Cardiff Royal Infirmary

Haemorrhoids

A friend once told me how pleased he was with his local Kupat Holim (equivalent of an HMO or Sick Fund) doctor.

He had had an attack of haemorrhoids (piles) and wanted some cream to relieve the itching. He went to his doctor and told her the story and asked for some cream which she gave him straight away with no fuss or bother. The cream helped him a lot and he was very satisfied.

"This was the first time you had haemorrhoids?" I asked.

"Yes," he replied.

"So, you knew they were haemorrhoids because you looked at them with a mirror."

"No," he laughed.

"But you looked in medical books or at least on the internet to see what haemorrhoids are?" I asked.

"No, that wasn't necessary, I know what haemorrhoids are."

Well, I don't blame him. But his doctor! How could she treat him without examining him? How did she know it wasn't a polyp, a rectal carcinoma, or rectal prolapse? Even if he did have haemorrhoids, how did she know it wasn't the presenting symptom of pathology higher up, WITHOUT EXAMINING HIM.

Abdominal pain

A similar incident occurred even closer to home. One of your relatives, Edward, had been having attacks of sharp upper abdominal pains on and off lasting a few seconds/minute or two over a period of several months. Other than that, he was in good health, no weight loss, no loss of appetite etc. However it disturbed him and he went to his doctor and told him this story. You might expect the next thing the doctor did would be to lie him down on the examination couch and feel his abdomen (and of course do a PR). But no, he sent him for blood tests (to this day I am unsure exactly what diagnosis he had in mind and which blood tests would have shown this).

These were normal so he sent him for an abdominal x-ray, which again came back normal. At this point he arranged for a gastroscopy. This procedure consists of passing a flexible fibre optic scope through the mouth and oesophagus into the stomach to search for and possibly take a biopsy of pathology of the stomach and first part of the small intestine. During this procedure the patient is fairly heavily sedated.

Although it didn't seem to me to be the most relevant test, at least I assumed he would now be seen and examined by a specialist gastroenterologist. I was wrong. It turns out family doctors have free access to list patients for gastroscopy without the patient having to go to the Out-patient clinic first; a very efficient system. So the first time the gastroenterologist would have seen Edward for an invasive procedure (complications such as perforation, although not frequent, do occur) would have been semi-comatose in the gastroscopy room!

At least he had the courtesy to see Edward after I rang and asked him to, but what about the rest of the population without this connection?

BE CAREFUL WHAT YOU SAY

Communication between people is always difficult. People use the same words to mean different things. Intonation, different cultures, different educational levels can all lead to misunderstanding. The following two letters, which did the rounds on the internet, do not tell a medical or personal story but I put them in as a wonderful, (and extremely funny), introduction to this problem.

Why men don't make good "Agony Aunts"

Dear Vincent.

I am a 33 year old woman with a successful career in software marketing. My husband was made redundant from his job as a store man in our local plastic party goods factory six months ago. I think he has always been resentful that I earned more than three times his salary, but since his dismissal he has become surly and very irritable towards me.

Anyway, as usual I left the house last week to go to work when about one kilometer down the road the car engine spluttered and stopped. I was completely unable to restart it, so I walked home to get my husband's help. When I arrived home I found him in our bed, fornicating with my next door neighbour!

When I discovered them, he had the cheek to say to me, "hang on; won't be a minute; nearly finished!" As you can imagine I had a screaming fit and told him (amongst other things) to pack his bags and get the h*ll out of my house.

After I had calmed downed a bit and started to think about things, I regretted saying this, as I still love him and I don't know what to do now.

Can you give me any advice?

Desperately yours,
Melinda

Dear Melinda,

It is difficult to give advice without knowing more details, but the commonest reasons for a stalled engine are :-

Dirt in the fuel supply line or blocking of the air intake manifold. If cleaning these doesn't solve the problem, check for excessive wear on the electrical brushes, or a leaky anti-drain back valve of the fuel pump. With the engine running listen for whistling or hissing sounds, which would indicate a leak in the vacuum lines. Alternatively there may be cracks in the carburetor diaphragm or nylon check ball.

I hope this solves the problem,

Vincent

Nowhere is this communication problem more acute than in the doctor/patient relationship. For some patients their doctors are the equivalent of the Regimental Sergeant Major or the Rebbe of ultraorthodox yeshiva students, or Imam in a Madrassa and the instructions are treated as imperative binding commandments.

Bedridden

Sometime in the early 1980's, a middle aged woman was admitted to University hospital Cardiff with an osteoporotic fracture of her femur.

She had wasting and contractures of her limbs, due to the fact she had been bedridden for over twenty years. When asked the reason for her invalided state, she told the following story. It was during the 'flu epidemic in the winter of one of the years in the 1960's. She lived in a small rural village in a remote part of Wales. She had called her doctor who came to see her in her home and told her to keep warm, drink plenty of tea and honey, take aspirin three times a day for three days and to stay in bed until he came to visit her again. Why he did not make this return visit is unknown; perhaps due to the pressure of the epidemic he forgot, perhaps he thought that as she did not contact him again, there were no problems and so there was no need, we will never know. But the fact remains, he did not make the follow up visit. Maybe she tried to get up a few days after the 'flu, felt weak and possibly a little dizzy and concluded the doctor must have been correct. Whatever the reason, she returned to bed where she stayed until some twenty years later when she sustained the fracture due to disuse atrophy.

Meniscectomy

Just outside Cardiff the orthopaedic hospital of Rhydlafer is (or rather was) situated. Here all the elective orthopaedic procedures were carried out. One afternoon on the pre-operative ward round of Professor McKay., we came to the bed of a young man by the name of David Jones, who had been admitted for a meniscectomy of his right knee.

Professor McKay went to examine him and asked him where exactly the pain was situated.

"Oh, my knee doesn't hurt me," he replied.

"Well, where was the pain when you injured your knee?"

"I haven't injured my knee."

"Look", said Professor McKay. "It says here in your notes you told me when I saw you in clinic you had twisted your knee playing rugby."

"No," he said.

"Well why did you tell me that in clinic?"

"I've never been to your clinic," he continued.

"If I didn't see you in clinic how have you come for this operation?"

"The hospital called me in."

It turned out that his next door neighbor was also called David Jones. He had, indeed, injured his knee playing rugby, had seen Professor McKay in clinic, had been diagnosed with a meniscal tear and was placed on the waiting list for a meniscectomy. Unfortunately the postman had delivered the postcard to the wrong house.

Does he take sugar?

When we were in Cardiff there was a weekly program on the radio entitled "Does he take sugar?" It was for and about disabled people. In addition to giving practical information as to what help was available and how to obtain it for various disabilities, it also explained in lay terms the various aspects of hearing loss, visual impairment, cerebral palsy, Down's syndrome and a host of other conditions to the general public and how these conditions affected those afflicted and their families. It was called "Does he take sugar?" because the lady who started and ran the program had a thirteen year old child with cerebral palsy. She had been invited, with her child, one day to tea with a neighbour. As they sat at the table for refreshments the hostess, after pouring a cup of tea for the child turned to his mother and asked her, "Does he take sugar?" To which the mother replied, "If you want to know if he takes sugar, ask him."

So many people ignore the patient, particularly if they are disabled or old and look at and talk to their parent/helper. Don't do it. If the situation is such that they are unable to answer, OK then there is no choice. But as the mother of the child in the above story was pointing out, these people are also human beings and entitled to all the courtesies you would give a non-infirmed person.

Interestingly enough, it is not only from the medical side of the table that this occurs.

As you know I am a children's orthopaedic surgeon and when I ask the child in my clinic what hurts him/her almost inevitably the mother answers for him or her. Obviously parental input is important, but I like to hear what the child feels. I point out to the mother I have the

"Original" in front of me and ask her to let him tell the story first. It is interesting how often the story differs, in particular the mother will almost always claim the pains are very severe and awful, whereas the child is often less concerned.

It is, perhaps, a rather nice reflection on Nature that the mother feels the symptoms far worse than her child! Here are a couple of examples.

Mother Love

Beautiful Baby

As you probably know a child, in his/her mother's eyes, is the most brilliant, handsomest, kindest, and most talented of all human beings ever to have walked on the face of the earth. Your sister told me this story.

Her friend Ofra had just given birth to her first child. It was during the first few minutes after birth as she was looking in bewilderment at this miracle of nature, this most beautiful, gorgeous perfect little being, unable to fully comprehend the sheer magnitude of exquisite beauty of her new daughter, that her doctor came into the room and saw her gazing totally engrossed at the newly born child. "Oh, don't worry," he said. "All babies are born ugly and she will improve markedly over the next few days."

Only a mother

I was in my clinic doing ultrasounds of babies' hip for DDH (dislocated hip joints at or shortly after birth). It is the wont of babies to cry. It's their language and the way they communicate their needs and feelings. All babies cry, each in their own distinctive way. A few do so at a volume that can be measured on the Richter scale, and a few with the pitch that can shatter granite at 100yds. The really talented can do both. It was such a sweetie that came to the clinic that particular afternoon. He was obviously hungry and cross about something, but we managed to finish the ultrasound. As I went to write the report, his mother turned towards me and with an idyllic smile, said to me. "He has such a musical voice, doesn't he doctor?" To my credit I just smiled, and said nothing.

Death threat

The Trauma centre in Cardiff was situated in the Cardiff Royal Infirmary and during my time there I heard this story. After his operation and having spent some time in the recovery unit, a patient was being transferred back to his ward- William Diamond ward. Awake but still groggy he overheard the following conversation between the nurse and the porter who was to take him there. In order to confirm he had the correct destination, the porter asked "Willie Dia?" to which the nurse answered "Yes, indeed."

The poor patient's shock at hearing this almost made it a self-fulfilling prophecy.

These things happen

The worst example of bad communication that happened to me was when I was "on the other side of the table."

It was your grandfather, Josef, who was the patient. He developed swelling in his left leg for no apparent reason and was admitted into a hospital in Haifa for investigation.

We were in the UK, but as luck would have it we had booked to come over to Israel at that time. Anyway, while in hospital they found he had a white cell count of over 80,000 (normal is up to about 12,000), grossly raised platelets and enlarged lymph nodes (the pressure on the femoral vein from one of which had caused the swelling in his leg).

A diagnosis of some form of myeloma was made but in order to know which form, a lymph node biopsy under local anaesthetic was decided on. For some reason the evening before the biopsy he had lost consciousness for a few minutes, but all was considered well enough on the day for the biopsy to go ahead. To this day I do not know exactly what happened but he died on the operating table. As you can imagine this was completely unexpected and a terrible shock. The junior "doctor" on the surgical team (I use quotation marks because this person, who, as will soon become clear, barely qualifies as a human being and certainly isn't entitled to the title of Doctor) was assigned to talk to me. In a monotonous rather bored voice he told me that your grandfather had died during the biopsy procedure. This was extremely difficult to take in and in a state of semi shock I asked him what had occurred. He shrugged his shoulders and said, "These things happen." What on earth did he mean 'these things happen'? Under his care, a fellow human being, a patient for whom he was responsible had just died while in his charge. At the very least, he should have cared a little bit, been upset, a modicum of distress, self-questioning, what had gone wrong? Why did MY patient die, was there anything else I could have done? Even if he didn't really care, he was talking to a relative whose father-in-law had just died and he should have shown some sympathy. But nothing; couldn't care less. Over thirty years have passed and whenever I think of him, it is with disdain, contempt and disgust.

In a similar vein, here are a few more examples, perhaps not so severe, but permanently burnt into the recipients' memories.

Sometimes it isn't what you actually say, but how you say it. I think these incidents occur when the patients are most vulnerable and a harsh or uncaring word can strike home with a far more stinging hurt than is intended.

Weight Loss

I was asked by an acquaintance, to look at her hip x-rays. She had been getting pain around the left thigh for some time and was under the care of an orthopaedic surgeon. She was grossly overweight. She was due to fly to London for a short trip later that day, and wanted me just to look at her x-rays. This is something I don't like doing as x-rays alone, as you are aware, are unreliable for diagnosis and only a small piece of the puzzle along with history, examination and any other relevant investigations. Anyway, I agreed, having explained to her the limitations of doing so. In the course of our conversation, she told me she had just seen a new orthopaedic surgeon, whose name she did not know. "I had to change orthopaedic surgeons because ten years ago I had trouble with my right knee and although it got better I will never go back to him."

"Why?" I asked her.

"Well," she said, "it wasn't so much the treatment, but he said to me 'as losing weight obviously isn't an option, you are going to have to live with this'. Who was he to say this to me? How dare he assume I am incapable of losing weight?" Ten years had passed and she still remembered his insensitive comment.

People are easily insulted

People can be very sensitive and can be easily upset, even if there is no hurt or criticism intended.

A friend's son-in-law, Jim is on a cardiology training program, which is hard work involving, like all specialist training programs, long hours. He and his wife, Eva, are very religious and are very committed members of the Orthodox community in Manchester. One day, Jim was rung up by the rabbi of their synagogue.

"Jim," said the Rabbi. "I am ringing because you haven't paid your membership bill."

Maybe it was because both he and Eva have busy jobs, four children under the age of six to look after, that they had overlooked to pay this bill. Anyway, as it happened, whatever the reason, the Rabbi was correct and their bill had been unpaid. Jim, however, was furious and put the phone down. He was so annoyed that they left that particular synagogue and joined a different one. What annoyed him was not that he had been reminded to pay the bill, but that it was the opening sentence of the conversation; not, "Jim, how are you?" "What are the kids up to?", "How is the job?" Etc., etc.

The Rabbi was right but sometimes it is better to be clever than right. As this example and that with our neighbour and Archie which is related further on demonstrate, it is the opening sentence or two that sets the tone for the conversation; be aggressive or nasty immediately and that is what you will get back; Action and Reaction. In this case the Rabbi lost a member of his congregation.

If you are going to complain or criticize someone or a firm, start with a pleasantry and if possible a word of praise, lull them onto your side and THEN give your complaint/criticism.

Ashkenazi/Sephardi

I was in the changing room of the gym and met someone I often see there. I don't know his name but we always talk a little. I don't know how we got on to the subject but he told me he was in the bank waiting in the queue, when someone just went to the front of the queue and started speaking to the bank teller. "How dare he, does he think I am like transparent glass, just non-existent? 'YOU PIECE OF RUBBISH' I shouted at him, 'How dare you, who do you think you are, get out at once.'"

"Anyway," he continued to tell me, "I threw him out." He must have recognized the shock on my face at the violence of his reaction, (he is a small quiet fellow).

He looked at me and continued, out of the blue. "You Anglo-Saxons were never in camps when you came to Israel were you? You never experienced the discrimination against us Sephardim. I remember my brother was in line and it was his turn and they gave the last job to an Ashkenazi."

Now I think the worse insult you can give someone is to ignore him/her or act as if they don't exist. If you are standing in a queue and someone just goes in front of you, it certainly is very anger provoking. However if he says, "Excuse me, I am in a terrible hurry, may I ask a big favour and go before you?" (time taken- 4seconds), you would let him with real pleasure (helping people is an endorphin releasing activity), i.e.-the total opposite reaction.

The other point of this story is that an insult or nastiness can stay for life and affect people's behaviour towards the insulter and even his group, religion, nationality etc. for years.

Status and Self-esteem

When I was sixteen, my father who was friends with one of the managers, got me a job working in the stores of a factory that made parts, under license, for the American Phantom fighter plane. It was quite an eye opener on the work ethics and practices in pre-Thatcher British factories, but the memory that stands out concerns the toilets. There were two sets, one on the second floor and one directly above this on the third floor. Those on the second floor were for the shop floor workers and those on the third were for the management. They were identical in all aspects; size, shape, colour, number of cubicles, basins, soaps, toilet paper etc. apart from one. Above the workers' toilets were written "Men" "Women" and above the management's toilets, "Gentlemen" "Ladies". It is said if you aren't a Liberal/Socialist at the age of 18 you have no heart and if you are still a Liberal/Socialist at the age of 28 you have no head. I am not sure I totally agree with the second part, but certainly, I was at that time, like many young people, a follower of equalitarianism, and I found the way the toilets were labeled rather repugnant. I asked my father's friend about this and he replied thus.

"Yes, I agree with you. However, when a worker from the shop floor is promoted to Foreman, amongst his new pay and conditions, is the right to use the management toilets, even though it means walking up an extra flight of stairs. *They* won't let us change the names."

Self-Esteem

One of the reasons I am writing this book for you, is because experience is the best teacher, but it takes a life time to obtain and thus often not available when you most need it. So I am hoping to pass on to you some of the things I have learnt over time which I wish I had known when I was younger.

One incident that occurred had been puzzling me for many years and it is only recently that I think I understand what was going on.

Unlike in the British system in which I trained, the hospitals in Israel do not specifically differentiate between the hospital workers. There is no "consultants only" car park. There are no "consultants only" dining room, no separate junior doctors' mess and staff canteen. There are no separate changing rooms in theatre for consultants, junior doctors, nursing staff and orderlies, only one car park for all the staff, patients and relatives, only one canteen and one changing room for all. I was having lunch in such a canteen when one of the hospital porters sat down at my table. He was nearly sixty, quite intelligent but poorly educated. In my pocket was a bright red biro which had been given to me by one of the drug "reps" with the name of the drug (as I recall one of the Ibuprofens) emblazoned on the side.

He looked at it and with a deep sense of injustice said "How come you were given a pen and not me?"

I had just come from the socially tiered UK system and was still adjusting to the more "socialist" approach, but I hadn't expected such a comment. I didn't know what to reply and muttered something like "Oh, I don't really know", but I remember thinking, "It's because I spent six years in University studying medicine, four years surgical apprenticeship and then another six years orthopaedic training. I treat patients with drugs such as this and the drug company thinks it worthwhile to advertise their products to the people who will prescribe them. You, on the other hand can't even read the (English) label on the packet, so why in heaven's name would any drug company in their right mind give you one of their pens?"

Now, as I said before his comment bothered me for years as I just couldn't understand it. Although he was poorly educated, he was by no means stupid and was very worldly. He had worked all his life in the hospital and was well acquainted with the medical hierarchy. There could be no way that he didn't understand why I got a pen and he didn't. I think that I now understand what he really was saying, was:- "Somebody has recognized your contribution and even given you a token of this recognition. While I may only be a porter, without me bringing patients to and from theatre, without me bringing in stores and restocking the selves, without me taking out the dirty linen and rubbish, there would be no operations and you doctors could do nothing. I, too, am worthy, but nobody recognizes it, and it's not fair."

And I think he was right. In our highly organized, companies, offices, organizations, cities, countries, societies there are so many "cogs" that must all mesh together for things to work. The big cogs get most of the financial rewards and social status. Those lower down (even the words "lower down" denotes inferior, less worthy- they are not physically lower!) on the social ladder get less. Those at the "bottom", the unskilled, the manual laborers get the least. Yet without them, nothing would work.

Now, I am not knocking the democratic, capitalist system. With all the injustices and inequalities this form of Society seems to work better than any other. But ALL the cogs are necessary.

So, what I am trying to tell you, if you get an opportunity to give a word of praise, a gesture of respect for the job well done to anyone "below" you, grab it and show them you appreciate them and let them know you think they are worthy human beings. Never, ever denigrate anybody, as in the next two stories.

Belittling people

There are many people around, perhaps because they themselves are insecure or have an inferiority complex, that have a need to show themselves to be superior by belittling other people.

You may know Dr. Harper. He is one of the best read and knowledgeable on more subjects than anybody I have ever met, and a good hour's edification and entertainment can be had in his company. Indeed, he is always very amenable to show off how clever he is. We were doing reserve army duty together, when during one of the meals, he accosted the cook, a simple hard working poorly educated man. He started to interrogate him as to why certain items which were listed as being on the menu were absent. It was as if he was a brilliant lawyer cross examining a lying witness on Law and Order.

A 19th century print of a barrister carrying out a cross examination

He totally destroyed everything this poor simple fellow said, in front of everybody, implying he was a thief and cheating us and the army. He was clearly enjoying entertaining everybody with his dominance over this far less articulate man. The more the cook squirmed the harder and sharper was Dr. Harper's attack. Eventually, the poor cook "lost it" and started screaming and shouting and had to be physically restrained from attacking Dr. Harper. It turned out he was holding back buying small items, so that at the end of our army duties he could buy more expensive items in order to give us a "slap-up" dinner.

Asking questions

There are several reasons for asking questions and encouraging people to ask them. The process involves thinking, analyzing what you have been told, looking for something missing, some inconsistency, something unclear. It is the very basis of intellectual activity. But just as the question is important, even more so is to whom it is addressed. It should be addressed to somebody who is likely to have the knowledge to answer it, otherwise, well, let the following story tell the tale.

We had been invited to friends for a holiday meal. Although not really orthodox, our host is "into" all the various rules and regulations, philosophy of Judaism. Towards the end of the meal the subject turned in this direction and he started quizzing me and, more so one of the younger guests, Tom, about what do students learn all the years in Yeshiva (religious seminaries). Patently we did not know but he gave no answers, merely repeated the questions or asked similar questions such as "What is the Mishna?" "What is the Oral Torah?"" etc.

Tom, not knowing, was shown to be more and more ignorant. Anyway we managed to turn the conversation away. But I was thinking about it later, and I should have said.

"Now, I know nothing about the history of Sweden, or the Philosophy of Kant. Of Chinese, Serbo-Croatian, Siamese, Sanskrit, Finnish, Burmese or Hakka I know not a single word. I am completely unaware of the intricacies of French medieval literature, the culture of the Mau-Mau, or the course of the Peloponnesian war. Fascinating as they are, Social Policies, Economics, archeology, Agriculture, politics are a bit of a blank to me. In spite of studying Science at school and Medicine at University, I have only minimalist knowledge of nuclear Physics, astronomy, botany, biology, organic chemistry. Mechanical electrical, civil Engineering, and car mechanics are a mystery to me. In spite of living in the Negev, I am totally ignorant about the life cycle of the Beer-Sheva fringe-fingered lizard, Nabatean social hierarchy, and run-off patterns in River Boker. I mention these as a few examples of the infinite number of subjects, out there in the Universe, of which I am extremely ignorant. I should, perhaps, have also included what goes on in a Yeshiva. Not because all these are not interesting. Indeed, NOTHING is uninteresting, but unfortunately life and our available spare time are so short we have to make choices about what to learn. What goes on in a Yeshiva is not of high enough priority for me to spend time on. However when I come across someone who *does* have knowledge of such an unknown subject, such as yourself, I hope he will be forthcoming with information about it, so I will be less ignorant afterwards. So my request is stop asking us questions to which you know we don't have the answer and tell us about it."

So when you ask questions do it because you want information and address them to someone with knowledge, and don't use your knowledge to ask someone questions just to make him look ignorant.

History

Up until the age of 16, the education system in the UK while I was at school was generalized and we had to learn a wide range of subjects and take anything from 8 to 14, "O" level exams. After that, the children were divided into different streams and took three subjects at "A" level (I took mathematics, physics and chemistry). Amongst the "O" level subjects was History, in which I initially was very interested. We had a boring teacher, although to be fair to him the syllabus he had to teach was also boring. We learnt lists of dates and names of the kings and queens of England and all the various wars and battles they fought. As is the wont of the bored child, my mind wandered to what it must have been like to live in those days, I wondered:- There was no electricity, so how did they see at night, did they have to go to bed at dusk? There were no fridges, so how did one keep food, particularly in winter when you couldn't gather a daily supply? How did people wash dishes, and themselves without running water (presumably not everybody lived directly by a stream)? What did they do without toilets, were there rules as to where to defaecate or could you just "go" where and when you wanted, did people care about privacy? And if we had to learn about kings, at least tell us what did they actually *do,* what was their daily routine? How long did it take a messenger to arrive from, say, Yorkshire, and surely by the time he had arrived the news was so out of date as not to be relevant. My mind continued to wander off and I tried to imagine what it must have been like to be in a battle. So there I was, clad in some leather tabard in a cold muddy field opposite hundreds of guys who were expending all their knowledge and energy trying to kill or maim me. I had managed to avoid the volleys of arrows but now huge numbers of people were coming at me swinging swords and pikes and axes. Doing the math's, being average in terms of physic and strength, I reckoned the odds of killing or being killed were against me. In that case, my next thought was, logically, "What on earth am I doing here?" I tried to fathom out, but I just could not work out, what motivated peasants, artisans, and landowners (whose role I was in) to leave their families, home and livelihoods to deliberately put themselves in a place where hundreds, even thousands of other people were doing their best to kill them. Perhaps even worse, having to kill someone they had never met, who had never done them any harm and probably if they were to meet in a pub, would be a great mate. I could and can understand why anybody would take up arms to defend his home, family and country from an invader- it is the best of two bad options, but these were the minority of cases. Most wars seemed to be family power struggles of the kings or leading barons (who spent most of the time ripping ME off and keeping me and my family in poverty while they lived in luxury) or some land squabble, or political or religious differences which, due to my total lack of education, I was completely ignorant, or because the particular monarch simply decided he wanted to go to war. So what I just couldn't understand and really wanted to know was:-

"Why did so many people, generation after generation, do it; go to war. What was in it for them?"

Scene of hand to hand fight at some unknown European battle by Hans Holbein the Younger (1497-1543)

What further bothered me was, like all kids I had had "physical altercations" with other children and knew how tiring this is. Even the strongest fittest professional boxers in the world can only fight for three minutes (and not even at maximum effort for these three minutes), before needing a break between rounds, so how long could an ordinary, probably malnourished, peasant, fighting for his life, maintain this intense physical effort? Logically, then, it seemed to me the whole battle (at least the hand to hand fighting part) couldn't last more than 1 or 2 minutes, but this was contrary to common sense. So I returned to the classroom, stuck my hand up and asked "Excuse me, sir, but how long did battles last?"

His sarcastic answer, of which he was obviously very proud and thought very clever, raising as it did, titters of laughter from the class was, "Forty-five minutes each way." Before I could pursue the question, the class comic, Mervin (who was a good friend of mine) raised his hand, "What about injury time, sir?" adding to the general amusement of all, except me as I really wanted to know. I gave up and returned to the battlefield. He went back to teaching us (or rather, to be more accurate, the rest of the class) about Wellington's campaign in the Iberian Peninsula war. Wellington, I wondered. How did *he* get to be the commander in chief? I could understand how lieutenants, captains, and majors could be promoted by the ranks above them, but how did Wellington get made the general in charge of the army. That he would prove to be a great general could only be known after his career, so who promoted him and why him. I put my hand up again "Excuse me, sir, but how did they decide who would be the general in charge?" At which point, Mervin shouted out, "They had General Elections." All the class INCLUDING the teacher laughed and I was very insulted and felt rather stupid. That did it for me and history. I closed my book and spent not one minute more learning or revising history and failed the "O" level (that taught this teacher a lesson!!). Times have changed and much more social history is now taught and I was able to find some excellent books on the life of middle ages peasants, although I still cannot fathom out why people so readily went to war.

Underestimating people

I failed History "O" level by a paltry 2 marks. Robert did it in style. Robert was a lovely fellow, but not (or so I thought) very bright. He languished in the lower half of the bottom stream in most subjects, including history, which was a bit odd as he liked history and wanted to take it at "A" level. In the childhood society I was raised, status was related to sporting and/or academic prowess. Robert was good at neither.

Anyway a few days after the exams were over we were relaxing and there was a school picnic to help us unwind. Robert was mooching around and I asked what the matter was. "I'm not coming," he said. The headmaster (who was also his history teacher) was keeping him in as a punishment and making him re-do the exam paper. "Why?" I asked.

"Well, it was a bit unlucky really. When the history exam papers were collected mine was on top and he read it before sending it off. I didn't know the answer to one of the questions and he got very cross with what I wrote and he is making me re write the exam instead of going to the picnic."

"What did you write?"

"Well, the question was 'Discuss British Foreign Policy towards Turkey 1895-1912' and I wrote 'They built a sweet factory and everybody was delighted.'"

47 years have passed and this still is, for me, the cleverest and funniest joke I have ever heard. And it was by Robert!! I was just blown away. Firstly to have the guts and audacity to write such a thing in the exam which will determine the rest of your life just blew my mind with admiration. Secondly, it was really clever (and it was by Robert!) and thirdly it was very very funny. I thought he deserved a really big prize rather than a punishment and couldn't understand the headmaster's reaction. From that day Robert holds my respect and admiration, and I am much slower to dismiss people as idiots or stupid; they can sometimes have great hidden depths.

Telling Lies

Lying is not just wrong, it doesn't make sense and rarely achieves it's objective.

What is, **IS**. Saying something that isn't "**IS**", is like building something on a vacuum. It doesn't exist. Furthermore, it is almost impossible to maintain a lie as one comment always leads to another, conversations flow and if deflected by a lie the ebb eventually gets so far from reality as to be an obvious fabrication. Once you get caught in a lie nobody ever believes you again. But you know this and what I want to write is about white lying.

Your girlfriend, wife, partner says to you, "Am I getting fat?"

Now all advice regarding relationships will tell you under NO circumstances should you say "Yes", in spite of the obvious and unmistakable four kilograms she has put on recently.

Maybe someone asks you a personal question and you don't want to tell him. Will you answer the truth "That's none of your business" or be gentler; "I'm not really sure at the moment what's happening?"

Certainly one can put up a very good case for lying in certain situations, but where does a white lie extend into something more significant? Sometimes when you think a white lie will spare someone's feeling, the opposite, as I will describe in the following story, results.

The school that I went to was a boarding school of about eight hundred pupils. We were divided into "Houses" and while we had almost daily contact with our housemaster, the headmaster was a distant and almost god-like figure. It was an excellent school and I loved every minute. Apart from academic subjects and sports, living closely with other people taught us how to get on with colleagues, friends, "enemies", how to give and take, compromise and stand up for yourself. Honesty, loyalty, consideration were instilled into us.

Every year there was an inter House Music competition. It was judged on how many people participated (one automatic point for every child in the concert), quality of playing, conducting, program chosen and other criteria I do not remember. However the automatic one point for every child who entered the competition meant anybody who could hold a musical instrument was expected to do his bit for his House and play something. I had "learned" the violin some years previously. However, I found holding the violin with extended arm and twisted wrist so uncomfortable I found it hard to concentrate on the music. I was further unable to consistently and accurately place my fingers in the strings, whilst simultaneous timing of quavers, semi-quavers was a total impossibility. My admiration and envy for musicians who can do so knows no bounds. However, I was co-opted to play in the concert which was attended by the judges and a variety of teachers including the Headmaster, and thus I obtained a House point. I neither rose to the occasion nor froze, but struggled through a short piece of simple music in my normal out of tune untalented manner. When I had finished, the Headmaster, who was sitting in the front row clapped heartily and said to me, with enthusiasm "Well done. That was really excellent." I was shocked. The first time the supposed beacon of truth, honesty, uprightness, decency, morality to whom we looked up, had ever spoken to me, he lied blatantly. When I got over the initial shock (even now it pains me), I was also exceedingly insulted. OK, maybe I couldn't play the violin but was I *so* tone deaf as to be unable to recognize an out of tune talentless performance? Was I *so* stupid I would believe white was black just because he had said so? Was I so devoid of any thinking ability to be so easily deceived into believing so obvious a falsehood?

Should I have played the game and said "thank you"? I am proud to say, I did not.

I said to him. "No, it wasn't." He was obviously embarrassed and didn't know how to continue the conversation. I remember feeling rather sorry for him, so I went on to explain, no matter how many times I practiced, all my effort and concentration went into reading each individual note on the page, translate it into where I had to put my finger on a string to play it and then actually make this difficult movement before doing exactly the same with the next note. This meant accurate timing was impossible, but what really amazed me was how musicians could remember thousands upon thousands of notes, the exact timing of each one, AND play them. He nodded in agreement and seemed interested in what I was saying, but who knows?

Charles Howard

My word is my Bond

I recently received a DVD called "Dignity:A Tale of two Wards." This is a 40 minute film made from the results of 16 observation teams, sent into different wards to record the interactions between staff and patients.

The chief researcher in this project points out that although there was no one totally terrible or one totally superb ward, for the purposes of this training film, the results were so divided into a bad and a good ward. I felt that this film should be shown in all nursing and medical schools. It was with this in mind that I contacted Professor Grey, Dean of the local medical school and arranged to meet with him and show him the film. For the first ten minutes of our meeting I explained about the film and Prof. Grey appeared very interested. After receiving a copy of the DVD, he said he would look at it and be in touch. The next twenty minutes, Prof. Grey spent talking about his own research into the relationship between markers found in leucocytes, rat brain, depression and anti-depressant agents. Whilst these maybe of some interest, they were neither appropriate or relevant to the discussion.

One month later, not having heard from him, I rang up his secretary and was eventually told that Prof. Grey had not yet looked at the DVD. A further month went by and again after ringing his secretary a few times, I was informed the film had not yet been seen. Realizing that Prof. Grey had no intention of watching the film, I spoke to the Dean, Prof. Kirby of an affiliated medical school. A copy of the DVD was requested and sent. A month went by; nothing. After ringing up his secretary, I was informed the disc had been lost and would I send a second copy, which was duly done. Several weeks went by and again, no contact. Again I rang up. "Oh yes," said Prof. Kirby, "I gave it to my deputy and if he is interested, he will contact you." I explained I needed to know his opinion as it would affect any further activity regarding the film. He agreed his deputy would be in touch. After a further six weeks and as contact still hadn't been made, I wrote a letter to both Deans, outlining this experience and ending the letter:-

"Please take the next suggestion in the spirit in which it is intended, but you would do well to watch this film, prior to returning it, as you are of greater need of its message than your students."

Within a few days, Prof. Kirby replied:

"I am writing to profoundly apologize for the delay in getting back to you. It is of course inexcusable, especially considering the initiative came from you and you have been so patient with us. I am afraid we have not the right slot for this CD, although I wish we did. I am going to mail it back to you and again on my behalf and that of the Faculty, please excuse our tardiness."

Prof. Grey is yet to reply.

I am putting this incident in because there are several important lessons to be learnt.

Firstly, if you give your word you will do something, do it! It is just as binding to say "I will do…" as it is to say "I promise I will do….", or "I swear a holy oath, on peril of my immortal soul that I will do …" or sign a document, witnessed by an attorney and 100 witnesses. Furthermore you must do it within a reasonable timescale. If, circumstances change and for good reason, you are unable to fulfill your promise, you must get in touch with the person and explain any delays, difficulties etc.

Secondly, as I will talk about later in the book, we all make mistakes all the time. If, as in this case, you forget to do as promised (although there were several reminders!), have the courtesy to apologize and do your best to correct your error (which often takes a lot more effort that the original task).

Thirdly, if you are on the receiving end of such errors and receive an apology, be gracious enough to accept it. Although I will be very reluctant to enter any future ventures with this faculty, I accepted the apology, and received the following reply:

"Thanks very much for your gracious response. I try to teach my children that one needs to know not only how to apologize, but just as important, how to accept one!"

Pardon?

The next "incident" occurred to your mother and I am sure you have heard her tell it on several occasions.

She had developed an erythematous rash on her face and had gone to see the local dermatologist, who is renowned for being rather antipathetic and aloof. She never smiles and hardly ever looks at the patient. Anyway she prescribed a cream which cost 100 shekels and didn't work (just stopping using her face creams allowed the lesions to heal up in a few days; but that is by the way). As you know Mum is getting a bit deaf particularly in one ear, and often needs to request people to repeat what they have said.

At the end of the consultation, this doctor said to her in a rather nasty tone. "You can't hear properly, why don't you get a hearing aid?" It is interesting that as people get older and need glasses, they have no problem, but for some reason having to wear a hearing aid (particularly for women) carries a stigma, and Mum was very insulted.

Perhaps if she had said, "I hope you don't mind me mentioning it, but I can't help noticing you are having some difficulty hearing what I am saying. Have you considered having a hearing test?" this story about an insensitive doctor would not have happened and would not have been told to all our friends.

Oh, yes, patients talk. Rest assured, very quickly, the population will know what sort of person and what sort of doctor you are!

Davening (praying)

We were recently visited by a friend, with whom I had grown up as a child in Coventry, and her husband, Anton. Your aunt keeps in touch with her, but I hadn't seen her for many years and we enjoyed a delightful reunion. During the course of their visit Anton told me the following story.

They were visiting the United States of America, when he developed septicaemia (a severe and very dangerous infection in the blood that can be carried to all organs of the body) and was admitted into the intensive care ward of a local hospital. He was exceedingly ill and it wasn't at all sure he would survive. It turned out he had a particularly virulent bacterial infection which wasn't sensitive to any of the antibiotics available in the hospital.

However, luckily for him, one of the doctors was very well read, kept up to date and remembered an article about a trial of a new antibiotic which was said to be active against this particular bug. This doctor spent a lot of effort chasing down the origin of the trial and asked for a supply of the drug. However it was by no means certain that it was available and that they would be able to get any.

"A lot of the time is very vague and I don't remember much of what went on." Anton told me. "But I do remember lying in bed with tubes coming from everywhere and asking what would happen if they were unable to get the drug."

"I'll never forget what one of the doctors said to me. He was head of department and I will always remember him with great affection. He said to me, 'Don't worry Anton, if we can't get hold of the drug I'll daven (pray) for you in Synagogue; works every time.'"

At this point in telling the story, Anton smiled gently to himself and repeated quietly, "Don't worry Anton if we can't get hold of the drug I'll daven for you in Synagogue; works every time."

Anyway, he continued, "As I said, I don't really remember much. There were a lot of doctors but I only remember him and one other. This one said to me 'You'll remember me Anton. I'm the one that saved you.' I didn't like him; arrogant fellow."

Head Injury

Perhaps the saddest example of communication failure occurred not in any of the hospitals at which I worked, indeed I do not know where or if this incident happened as I was told it and did not witness it myself. It wasn't what was said but what wasn't that was the problem.

As you know head injuries are fairly common and it was the custom of the day (is it still?) to admit all head injuries who had a history of loss of consciousness overnight for observation. Their vital signs are assessed every 15 minutes in order to pick up intracranial bleeding. The standard method was to put the patient on a head injury chart measuring blood pressure, pulse, response to stimuli and pupil size. One such patient, a young man who had fallen from his motor bike, was admitted to such a ward one evening. He had been examined by the junior doctor but apart from complaining of a mild headache and minor grazes, there were no other apparent injuries. He was sent to the ward with instructions to put him on a head injury chart, and this was duly done. Due to it being night time and the usual staff shortages, a first year nurse was assigned to this task and she did it admirably. She recorded accurately how one pupil started to dilate, how his pulse and blood pressure began to drop, how the second pupil started to dilate, how he initially became increasingly agitated and unco-operative, but later on became much quieter and was sleeping peacefully. Indeed she recorded all his signs up to and until the moment of death.

Nobody had told her why she was making these observations and when to call the on-call surgeon.

Two of the next three stories are not related to medicine, but I include them because I learnt some really important aspects of life and relationships from them.

Tomatoes

The first occurred in the supermarket, where I was doing the weekly shopping. Sometimes I leave the trolley and go off to collect some items and bring them back to where I had left the cart. I have noticed a lot of people do this. Anyway, I was returning with whatever it was that I had collected and at the other end of the aisle I saw my trolley and a man taking out my packet of tomatoes. What a cheek! I went up to him and confronted him rather aggressively, "Excuse me, but this is my trolley, get your own tomatoes" or a similar such comment; I don't remember my exact words.

"Oh, I am most terribly sorry," he said in a gentle quiet voice, "Actually these are my tomatoes and as you can see your trolley and mine, which is over there, contain very similar items. I mistakenly put my tomatoes in your trolley. As you can see your tomatoes are there (pointing to the bag in my trolley). I am terribly sorry, it's my mistake." Well, that was true, the mistake was his, but his soft patient way of explaining the matter left ME groveling and apologizing profusely (and rightly so).

Supermarket trolley

The next story is in a similar vein and you too were involved.

Archie

I don't know if you remember the next incident, but you were actually present. It occurred in our street. I can't remember the exact details, but Archie (our dog) had been a bit of a pain. I am not sure if it was because of barking growling or chasing cats or other dogs, but he had managed to annoy one of the neighbours across the street. We met this neighbour in the street and in a slightly aggressive annoyed tone he started to complain. Now, to be fair to him, he actually was in the right, but if people come at you in an aggressive manner, the normal (but incorrect) reaction is to be aggressive back, and so we jumped to Archie's defense and were equally unpleasant. Shortly afterwards, it became known that this neighbour was suffering from metastatic prostate cancer, which can cause horrible severe bone pains, and within three months he had died. I was sorry afterwards that I had spoken to him in such a nasty way, but I was too late to apologize.

Author's portrait of Archie

There is an old American Indian saying: "Don't judge a man until you have walked a mile in his moccasins." So when people are rude, inefficient, unhelpful, I imagine the following scenario.

He got out of bed in the morning to the usual nagging of his wife; why did he always walk on the carpet with his dirty shoes, why hadn't he taken the dustbin out as she had told him, and not to forget to buy milk on his way home *again*. On the way to work he got a speeding ticket which made him late and was shouted at by his boss, the coffee machine was broken but he only saw the notice after he put his money in the slot, while going up the stairs to his office he tripped and got a large painful graze and bruise which elicited much laughter from his colleagues. And so on and so on,-- you get the picture.

Old daughter

While I was in Swansea, one of the orthopaedic surgeons I was working for was Mr. Brown. As you know as the number of old people is increasing and so is the number of osteoporotic hip fractures. The treatment of choice is to operate on these fractures and get the patients mobile again as soon as possible. Not infrequently these fractures tip a barely mobile fragile old lady (most of these fractures are in elderly women) into a chair bound person who can no longer look after themselves. There is a great shortage of homes in the community for these poor people and it is often difficult to move them from an acute orthopaedic bed out into some form of rehabilitation or other accommodation. It is this epidemic that is one of the greatest medico-social challenges of modern western society. Anyway we had admitted a ninety five year old lady with such a fracture. She had been living with her daughter and son in law until the time of her fracture. She had multiple physical and mental problems concurrent with her advanced age. After her operation she progressed very slowly. That is to say she survived the operation and was able to sit out of bed in moderate comfort but was unable to walk even with a walker and a helping physiotherapist. Nevertheless, the pressure on the trauma beds was so great we wanted to send her home with support from home physiotherapy and social services when we were informed by her family that they were unable to cope with her at home and refused to take her back. Mr. Brown was outraged and infuriated by this. The thought of someone abandoning their mother was an anathema to him and he hit the roof when he heard about this. "Make them an appointment to see me," he fumed. "This mother brought them up, looked after them when they needed her and now they think they are just going to dump her like a piece of rubbish. I don't think so."

The appointment was duly made and the thought of this daughter not wanting her mother back was rankling and he was working himself up to give her a good piece of his mind. The appointed hour arrived and in walked, or rather hobbled a little old lady of 75, hands knobbled with arthritis, back bent from osteoporotic fractures, knees stiff and degenerative. She smiled delightfully at Mr. Brown and said, "You wanted to see me about Mother?"

To his credit, and I was very impressed by his quick thinking, he rapidly reassessed the situation. "Yes," he said. "I just wanted to let you know how pleased we all are at the progress your mother is making and we hope to get the social services to find her some sheltered accommodation or a residential home as soon as possible."

Moral of the story. Never get angry with someone without knowing all the facts; (better still- never get angry).

The other side of the table

Being a patient oneself is always an eye opening experience. In 1988 Edward Rosenbaum wrote a book "A Taste of My Own Medicine: When the Doctor Is the Patient" which was later turned into a film "The Doctor" starring William Hurt. The doctor, having been misdiagnosed initially, is treated for throat cancer. He experiences what it is like on the other side of the bed. Other than recommending that you see the film or read the book, I'll say no more about this book/film, but I will tell you my experience. Although perhaps trivial, it is none the less poignant for having actually occurred.

Apart from hypertension, which I have had since my early thirties, and is well controlled, I have been very lucky to enjoy good health. It was only because I wanted a health certificate for the gym, that my GP sent me for an exercise (treadmill) cardiac stress ECG and an Echocardiogram.

The ECG was carried out by a young female technician. When I came into the room, she told me to take my shirt off and take one step forward. This was a little difficult because there was a dustbin just in front of me, but I did so and without further a word she took a razor and started shaving off some of the hairs from my chest (which fell into the dustbin) and to stick on electrodes. Would it have been too difficult to utter one sentence of explanation, "I'm just going to shave off a little bit of your chest hair so I can stick on these electrodes," before assaulting me? I don't think so. Her entire demeanor was as if she was processing meat or working on some routine job which did not require the effort of communication. During the entire procedure no explanation of what was going to happen and she only uttered the minimal word here and there as was essential.

The ECHO procedure was carried out by a technician and a cardiologist, whom I have known for over twenty years. He was busy writing up the previous patient's report when I entered the room, so I did not disturb him. The technician started the test (again no explanation that she was going to put some cold gel on my chest and image the heart by rubbing a transducer over the chest wall, and that I would feel nothing. One sentence, that's all it would have taken). Anyway they started discussing the findings and it was only half way through the examination, when I made some comment that the cardiologist looked up and said. "Oh, hello Charles, I didn't realize it was you." Of course he didn't. He hadn't had the common courtesy of looking up, seeing with whom he was about to interact and saying a word or two as I came in the door. All he was interested in was the pictures on the television monitor of my heart echo. If nothing else, it was just downright rude. Had I been a non-medical person (as are the vast majority of the patients) who was totally unaware of what the procedure entailed, it would have been quite a harrowing and worrying experience.

It is this deliberate lack of communication, failure to make the connection between person to person that is most harrowing to patients. Many patients complain that since the introduction of computers the doctor doesn't look at them anymore, just at the screen. This lack of eye contact distances the doctor from the patient. The story that follows, although about nursing staff, demonstrates this quite well.

Loss of caring in our profession

In all human activity mistakes happen all the time. A good organization recognizes these mistakes and puts into place checks and routines to minimize and correct errors. Errors in giving out of drugs on the ward drug rounds occur frequently. Over 2000 serious complications per year are thought to happen because of wrongly administered tablets and medicines, which means there are vastly more which are unreported as the error caused such minor effects that they went undetected. One reason for these mistakes is the distraction of the drug nurse's attention by various interruptions/questions from the patients. In order to overcome this, nurses were given tabards to wear with the words; "Do not Disturb. Drug round in progress" and a three month trial showed that, indeed, the number of interruptions was reduced as was the number of wrong medications.

The above paragraph certainly fits with the practice of the Good organization to a tee. But the general feeling that doctors and nurses are becoming increasingly impersonal and remote and that this would even further distance the nurses from the patient caused a general outcry (and rightly so). This proved so effective that wearing of these tabards was stopped.

What is so instructive about this is that these tabards were clearly shown to be effective in reducing medication errors and save lives and decreased illness, but the price, whether just perceived or real, of the loss of carer contact was just too high.

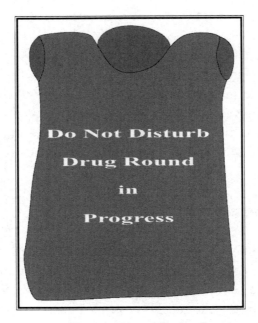

A "Do not Disturb" tabard

PATIENTS ARE OPTIMISTIC

She will get better, won't she?

Sometimes though, no matter what you tell the patient, no matter how hard you try to explain things to them, they won't accept what you tell them.

While I was a junior house officer in general surgery at Lewisham hospital, a middle aged woman was admitted having sustained a cut on her foot two days previously, which had got infected. She suffered from a very severe form of Systemic Lupus Erythematous (SLE). She was extremely toxic, in renal shut down (no kidney function), and only semi-conscious. Microscopic examination of pus from the wound showed she had necrotizing fasciitis. The infection together with her severe peripheral vascular impairment (very poor blood supply) due to the SLE, had caused septic shock and gangrene in the lower limb. So severe was the gangrene that within a few hours of admission there was visible deterioration and the gangrene extended above the knee and was continuing to progress rapidly.

There was very little hope of recovery; antibiotics couldn't reach the infection due to destruction of the blood vessels. Her only hope was a high above knee amputation, massive antibiotics, dialysis and respiratory support. Her only living relative was her unmarried daughter who was borderline mentally retarded and lived with and was looked after by her mother. It was my job to explain to her the situation and get a consent form signed for the operation. How do you tell this poor girl her mother, her only support, was almost certainly going to die?

I started off as gentle as I could.

I told her, "You know your mother is very very ill."

"Yes, doctor," she replied.

"Obviously, we are going to give her all the best treatment and antibiotics, but unfortunately she is going to need an operation."

"OK, doctor."

"We hope she will recover, but it is a very severe infection and because of her previous illness, unfortunately she has a lot of very difficult other problems. Do you understand what I am trying to tell you?"

"Yes, doctor." Patently, she didn't.

"She really is very very ill and we will do our best for her, but I am not sure really what is going to happen."

"OK, doctor."

"We are hoping she will recover, but she is so ill we are not sure. She needs a very big operation and unfortunately we will have to remove her severely infected leg which is sending infection into the rest of her and we need you to sign the consent form."

"OK, doctor no problem. Where do I sign?"

From her manner it was obvious she didn't have a clue about the severity of her mother's condition.

I said to her, "There is a very big possibility she isn't going to survive the operation."

At this point, she burst into tears. Obviously I tried to comfort her as much as possible, but to tell the truth I was relieved that at last she had at least understood the situation and hopefully it wouldn't be so much of a shock a few hours later when her mother would inevitably die.

I turned to leave the room and as I was going through the door, she stopped crying, sniffled a bit and through her sniffles said, "But she is going to get better, isn't she?"

During my training my professor once said to me that people can cope very well with deformity and disability, what they can't cope with is chronic pain. Here are two examples of what he meant.

"'oovering"

When I was in Swansea I was working with a hand surgeon, Mr. Neils. We saw in clinic a patient, a Londoner with a wonderful Cockney accent, who had undergone four operations for severe Dupuytren's contractures of the palmar fascia. This condition can vary from a small dimple in the palm to contractures so severe that the fingers dig into the palm and can't be extended. This patient was one of the mostly severely affected and in spite of complete removal of the contractures at each operation followed by long courses of physiotherapy, the contractures had returned and he was only able to move his little and ring fingers slightly. It was a very disappointing result and Mr. Neils said," I am really sorry we started any surgery. After all the effort, time and resources, you are no better off than before."

"Oh no, doc", replied the patient. "Things are much better now."

Much amazed, Mr. Neils asked him what exactly he meant.

"Well," he said. "I am retired and I don't have any active hobbies, but I do like to help the trouble and strife (wife) around the house. Before the operations I couldn't do anything, but now I can do a bit of 'oovering."

Spinal injury

This story was told to me by Mr. Yona, a consultant in Cardiff.

When he was doing his orthopaedic training just after WW11, he was sent for three months to the new spinal injury unit at Stoke Mandeville Hospital.

One of the patients was a sailor who had fallen from the Crow's nest (ship's observation post high upon one of the masts) and sustained a very high cervical vertebral fracture which had caused complete paralysis of his body below the neck. He had no use of his arms or legs, was doubly incontinent, and was confined to a spinal bed. He had no family and friends had long since stopped coming to visit him. His daily routine was unvaried and consisted of meals, two hourly turning to prevent pressure sores, doctor's ward round and a half hour daily physiotherapy session to prevent contractures.

As prism glasses were not yet available, he was unable to read a newspaper or watch television etc. His view of the world was the ceiling or the floor. Nevertheless he remained in good spirits and was always happy to see the doctors on their daily ward round.

Prism glasses being used to read a book lying down

One day Mr. Yona said to him. "I hope you don't mind me asking, but before coming here you were an active lively young chap with all your life before you. Suddenly, one fall completely changed your life. Now you are reduced to this monotonous regime totally confined to bed. How is it you manage to keep so cheerful?"

"Well," came the reply, "It could have been much worse. I might have been killed."

The lookout post high on ship's mast known as a crow's nest[6]

---

[6] http://commons.wikimedia.org/wiki/File:Tarmo_crow's_nest_Vellamo.JPG License Public Domain **Photographer MKFI**

GRATEFUL PATIENTS

It may be surprising but although patients may be grateful for what you do for them, they are far, far more moved by the way you do it and how you show them you care; even if the treatment is unsuccessful, and you have "given" more than standard ordinary treatment e.g. rung them at home to inquire how they are, rung around other doctors to ensure they get seen by the expert you want etc. I first came across this, although I didn't understand it for many years, again during my time in Lewisham.

A tin of Biscuits

A forty five year old woman was admitted with a perforated gastric ulcer. The registrar, Mr. Leach, saw her in the accident and emergency department, diagnosed the problem and immediately arranged for her to go to surgery. She did, indeed, have a 'perf.' and the start of peritonitis (severe inflammation of the abdomen from the leaking stomach contents). She was in a moderate to severe condition but gradually over the next two weeks with constant care, "drip and suck", abdominal drains, electrolyte monitoring etc she made a good recovery and was about to go home when she collapsed in the toilet. Luckily for her, Mr. Leach was on the ward at the time and he resuscitated her, made the correct diagnosis of a pulmonary embolus (a blood clot from one of the large veins that breaks off and travels to the heart and then the lungs, blocking further blood flow). He instigated the correct anticoagulant therapy. As you know looking after these patients is a lot of work; constant monitoring of blood gases, electrolyte balance, respiratory function etc. etc. She improved and again was about to go home when she spiked a small temperature and complained of mild abdominal pains. Mr. Leach was unhappy and decided to keep her in a little longer. The next day the pain got worse and he made the diagnosis of an acute appendicitis, again operated on her and at operation found she did indeed have an inflamed appendix. Over the next few days she again improved, and to everybody's relief, she developed no further complications and went home after roughly two months in hospital. I next saw her about six weeks later going up the stairs to her old ward. She had just been seen by Mr.Leach in the clinic, given a clean bill of health and discharged. When I saw her I said hello and asked her how she was etc. In the course of the conversation she told me she was just on her way up to the ward to bring the nurses a tin of biscuits. To Mr. Leach who had saved her life three times, spent many hours (as indeed I also had) making sure all was going well with her treatment, she hadn't even said good-bye when she was discharged. At the time this was a great shock to me and incomprehensible for many years. Now I think I can understand it. From her prospective, the doctors were distant almost "Alien" beings rushing around discussing her "case' at the end of the bed, doing painful and unpleasant things to her. She saw us day and night (in those days we did 1 in 2, thirty six hours on, then the evening and night off, except for the weekends when we would be on from Thursday morning to Monday night, or Friday morning to Tuesday night, every other weekend). In spite of having these long hours available, the pressure of having over sixty surgical patients to look after, covering the Accident and Emergency room, emergency operations, elective admissions, operating sessions, ward rounds and outpatient work, and study for F.R.C.S., meant there was little time left for "social interaction" with our patients. We were distant figures, but it was the nurses who straightened out her pillow when she was uncomfortable, it was the nurses who brought her the bed pan to relieve her or a cup of tea at one O'clock in the morning when she couldn't sleep. It was the nurses who gave her a painkiller to ease the pain of the operations. Thus it was to the nurses that went the tin of biscuits.

A tin of biscuits

The next story may seem at first sight to be the opposite, but actually it is again another example of, "it is not what you do, but how you do it".

Back Pain

This story concerns the consultant orthopaedic surgeon, whom I have mentioned before, Mr. Simmons. He was a man of limitless energy, great charm and charisma. He had a huge private practice but unlike some of the other consultants, still gave 100% and more to the NHS. Unfortunately, although he was, on the whole a very competent and technically excellent surgeon, his clinical judgment was terrible, bordering on nonexistent. He said once to us, "I am a surgeon, so if people come to me it is because they want an operation. I am here to provide a service and so I always operate on them."

Amongst other things he operated on backs. The problem with back surgery is the patients. Everybody gets attacks of back pain during their life but patients with non-resolving back pain form a special category. Some will have genuine pathology, disc prolapse, stenosis, infections tumours etc. but a large proportion have some supra-tentorial overlay. Whether it is to get off work, because they are overweight, do no sport and have poor musculature, there is no doubt the back symptoms are often secondary to other pathology. Anyway, Mr. Simmons tended to operate on all of them and then they were followed up in the back clinic. He had a rule that because there were so many he could not see everybody, but would see them every third visit. We all knew this and so did the patients who came year after year (sometimes several times a year). It was a nightmare. Each patient had a file as thick as a fist and was totally disinterested in seeing the registrar.

A typical conversation would go like this.

"Hello Mr. Jones how have you been recently"

"Terrible, I am much worse. I am in constant agony and I can't walk far. Any lifting of even small weights sets me back off and as for sitting, after two minutes I get completely stuck."

After examining the patient complete with "ows" and "ouchs" apart from forward flexion of a few degrees, SLR (straight leg raising) of 5-10 degrees (i.e. much less than occurs in a genuine disc prolapse), the scars of various laminectomies, discectomies, fusions etc, there would be nothing new or specific that would be of much help in treatment. Often there was a much greater degree of active movement on and off the examination couch.

"OK Mr. Jones, let's try some physiotherapy."

"I had that 6 months ago. It just made me worse."

"Alright, maybe we will try you on a course of Naxyn."

"Look, doctor, if you bothered to read my notes you would see I was given them two years ago and they brought me out in a rash. I am absolutely forbidden ever to take them again. They could kill me. Can I see Mr. Simmons, please?"

Anyway, one day, I and another registrar, Tony, were doing this clinic which we all dreaded, when a woman in her forties was brought in in a wheel chair. She had had four operations, each one making her worse and she was now totally wheel-chair bound. Tony looked through her notes to see what her problem was. It turned out she had had nonspecific back pain and before she had seen Mr. Simmons, she had been to three other consultants, all of whom, finding no gross pathology, had discharged her back to her GP with various tablets. She had even had a psychological examination and had a number of varying psychological problems which were felt to make her dependent and there was an element of conversion of her psychological problems into the back pain. However after four operations, each with increasing scar tissue and instability, she now had genuine weakness and sensory loss in her lower limbs and, if my memory serves me correctly, she may even have had continence problems. Tony said to her "I'll bet you rue the day that you first saw Mr. Simmons." Agreed this was not a particularly professional or appropriate comment from a junior member of the team to say about his boss, but it was her reaction that was so remarkable.

I think had she been able to, she would have jumped from her wheel chair and throttled Tony.

"How dare you say that!! How dare you. You must know Mr. Simmons was the only one who tried to help me."

The woman was actually grateful to him!

Mistakes

This incident occurred in the operating suite in my hospital. The head nurse, whom everybody liked, was very efficient and always helpful, had, by mistake, booked my list and the plastic surgeon's list into the same operating room at the same time. We normally alternated weeks, but this week we were due to operate on the same day. She had been informed but, as it turned out she hadn't made the correct arrangements. The plastic surgeon was absolutely furious and started shouting and threatening to go to the hospital manager and this just wasn't good enough etc. Without wishing to be conceited, just as I am ashamed of how I spoke to our neighbour, I am pleased with how I reacted.

"OK." I said, "We have a problem. How can we sort it out? Maybe the anaesthetist who was on call and due to go off would stay for a few extra hours as a one off and we can share a nurse and open up the day surgery theatre?"

Unfortunately this proved impractical and after several other suggestions, we had to cancel one of the lists. This is very traumatic for patients (and their surgeons) who have been admitted to hospital, organizing their work and holiday schedules to fit, are fasting and are "psyched up" only to be told, "sorry we can't operate on you today, you'll have to go through all this again at a later date".

In the end, I think we had half a list each.

Afterwards the head nurse was very grateful to me, which I found embarrassing as it was unnecessary as I hadn't actually helped very much. It was then that the following sentence came into my head; I don't know where from.

"No, you mustn't thank me. It was purely self-interest," (although actually I am rather proud to say it wasn't!).

"We all make mistakes. If you show me a man who claims he doesn't, I'll show you a liar. Today it was your turn; tomorrow it will be his or mine. And when we do make the inevitable cock-up, who will you try and help, him or me?" She just smiled and said nothing.

This incident set me thinking and was the origin of my (I do not claim it as original) "philosophy" which you will have heard many times.

"Don't judge a man by his mistakes, but how he tries to correct them."

OUR WORST NIGHTMARE

You have finished examining your patient and have come to a certain conclusion; a diagnosis. It turns out you were wrong. We are all fallible and make mistakes every day. Often it may not matter that you got the diagnosis wrong as the vast majority of conditions get better by themselves or would have required the same treatment anyway. If the patient has tendinitis or bursitis you may well have prescribed anti-inflammatory drugs with good effect. If you tell a patient with knee pain from overuse to reduce or stop activity for a short while, perhaps it doesn't matter that you thought the pain was arising from the patella when in fact it was from the patellar tendon.

But often it does matter and there is nothing the newspapers enjoy more than reporting on a case that ends in tragedy due to a missed diagnosis or misdiagnosis. A Google search of "missed diagnosis" gave over thirty one million hits. Here are a couple of personal examples.

Elbow fracture

As I am sure you remember, your friend Matan fell on his arm while playing football. He was seen by an orthopaedic surgeon and x-rays of the wrist, forearm and elbow (AP lateral and obliques) taken. Matan said the doctor didn't look at the x-rays but read the radiologist's report which was as follows:- "X-rays forearm and elbow including obliques. Intra-articular undisplaced fracture of radial head. Additional supracondylar fracture. Intra-articular haematoma."

So You're a Doctor Now

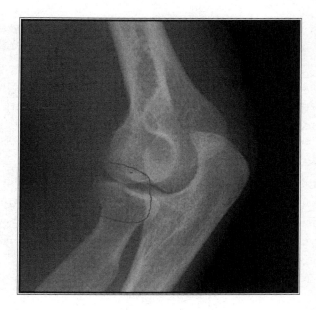

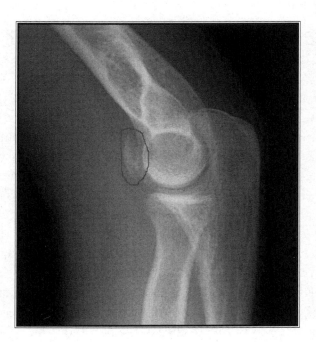

Matan's x-ray showing the undisplaced intra-articular fracture of the radial head and displaced fragment of the capitellum.

As you can see from these x-rays, there is no supracondylar fracture but there is a displaced fracture of the capitellum. This can be a serious injury. Left untreated, the fragment may cause locking of the elbow in the short term and arthritis of the elbow (and wrist problems from proximal migration of the radius if the fragment is large). These fragments should be reduced and pinned. Small fragments should be removed.

Also note there is ABSOLUTELY no difference in the soft tissues of the arm/forearm and joint in these x-rays. The radiologist knew, intra-articular fractures bleed into the joint, therefore he diagnosed a haematoma in the joint. Beware of finding physical signs because you know "they must be there"! Be objective when examining patients/radiographs etc.

Subluxing patella

An example of missed diagnosis happened to me fairly recently. A mother brought an eleven year old girl to clinic. I had seen her four years previously at which time she had complained that she tended to fall because her right knee gave way. From my notes at the time, I recorded she had a normal gait, full range of pain free movement of her hips knees and ankles. There was no ligamentous instability of the knee. She could tip-toe and heel walk and there was no muscle wasting. Gower's test (for muscle weakness) was negative and there was no deformity or neurological deficit. I reassured them and told them to return as necessary. Well, the symptoms improved although I gathered at the next clinic visit four years later, she did still get this giving way occasionally. However over the previous three or four months the frequency had increased and was beginning to bother her. What was immediately obvious was that she had a very slight, but definite thickening of the knee synovium. Again all the signs were negative until I asked her to sit on the edge of the bed and flex and extend the knee, whereupon it became immediately obvious that the patella was maltracking.

So You're a Doctor Now

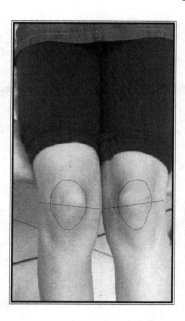

Both legs are show with the knees almost fully extended. The patellae (knee caps) are outlined and can be seen to lie roughly in the middle of the femur.

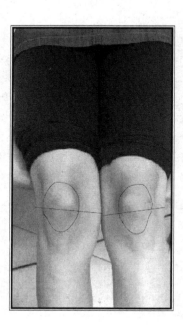

The knees are now fully extended and the right patella can be seen to have moved markedly laterally (towards the outer aspect of the knee) as compared with the left (normal) knee. A bit more lateral force by a knee twist and the patella can completely dislocate laterally.

Was this sign present when she was seven? Maybe not, but I suspect it was and I missed it. I had been concentrating on muscular or neurological conditions and maybe I didn't check for this particular condition.

The medical profession takes this seriously and there are research papers in the literature. A recent study[7] from America estimates that there at least 150,000 people per year, die or are disabled as a result of misdiagnosis in office consultations. In addition to this there must be many, many times this number that produce no serious harm; for example all the viral conditions treated with antibiotics. But diagnosis can be very difficult and the signs and symptoms of, for example, a viral sore throat and a streptococcal infection can be identical.

So why does it happen and what advice can I give you so it doesn't happen to you? I think it occurs for a number of reasons. Firstly, diagnosis can be very very difficult, particularly in the thousands of less common conditions which you have not come across before. If you are unaware of the condition (and there are so many it is impossible to know them all even within your own specialty), you won't think of it.

Secondly although a good history is vital, it may be misleading. The patient will tell you what he/she thinks is important. For example, a patient tells you every time he eats, he gets back pain. This may lead you into spending your limited time with him investigating the gastrointestinal tract. Does he have an ulcer, duodenitis, pancreatic tumour, gall stones? But what actually is happening, is that when he

---

[7] Measuring Diagnostic Errors in Primary Care. The First Step on a Path Forward Comment on "Types and Origins of Diagnostic Errors in Primary Care Settings" David E. Newman-Toker, MD, PhD; Martin A. Makary, MD, MPH JAMA Intern Med. 2013;173(6):425-426.

sits at the kitchen table on a rather low chair, the pressure in a central lumbar disc prolapse increases and was the cause of his pain. It was the sitting not the eating that was the cause, but in the patient's mind the association was with food, and thus he directs his tale towards a gastrological condition. His subjective view point will even affect his answers of direct questions which you may ask him. In clinics and the emergency rooms there are always another four or five patients waiting to be seen and time is very limited. Maybe if we had an hour with each patient, missed diagnosis would be much rarer. But you don't have unlimited time and you have to follow where the history and examination lead you.

Thirdly, it can sometimes be very difficult to see the changes in the body produced by the disease, particularly when it has just started.

So how can we reduce the incidence of missed/wrong diagnoses? Well, to start with be really thorough with the history examination and investigation. The vast majority of conditions get better within a known time span (colds/flu a few days/week, back pain up to 6-8 weeks etc.). So if patients aren't improving and don't fit the expected path, bells should ring. An extreme example is the sad case of Jean Cross. She had pain in her left shoulder and visited her GP thirty nine times over a period of three years. She was given repeated prescriptions of pain killers. When she was eventually given an MRI scan a huge lung tumor pressing on the nerves to her arm was found and a few days later she died. Thirty nine visits over three years!!

Take every opportunity to re-examine your patients. There is a tendency to think "I examined them yesterday; nothing will have changed so I won't find anything different" (I presume this was the mistake made by Mrs. Cross's doctors). Maybe you won't. Maybe you were right first time and there is nothing pathologically wrong. Maybe you'll do the same examination and miss the signs again. BUT, give yourself and your patient the chance that maybe, you will find something new which will help in diagnosis and treatment.

Again, examine, re-examine and re-examine at every opportunity.

Charles Howard

Sometimes you just need to be lucky

When I was working in Soroka Hospital, I was called to the Accident and Emergency room to see a young soldier with pain and swelling in his knee. I forget his name, but he was a 20 yr. old American volunteer, who was in the Parachute regiment. He had no family here and was part of the "Lone Soldier" group of very positive Zionist youth who come here at his age. He had completed some grueling strenuous exercise the day before and several hours later developed pain and difficulty weight bearing on his left knee. To reduce the possible waiting time, the Emergency room staff, knowing how long it sometimes takes for us to arrive in the A&E, had a habit of calling two or if possible three doctors to get an opinion. In this case they had also called the rheumatologist, Dr. Freed, and he had arrived about fifteen minutes before I did. He had examined the patient and noted that he had an effusion (fluid) in his knee. The patient was apyrexial (no temperature) and the rest of the examination was unremarkable other than a hobbling gait due to his painful knee. Nevertheless, he sent off blood examination and did an aspiration of the knee which produced clear synovial fluid. While waiting for the results of the blood tests, I arrived and on examination confirmed Dr. Freed's findings with no further additions. The blood results came back normal and examination of the synovial fluid showed nothing remarkable. I had written the discharge letter giving him "Betim"; (rest in camp) as he had no home in Israel to go back to, and anti-inflammatory drugs. When I finished the letter, the Emergency room nurse told me "Oh, no, Dr. Freed has admitted him." This seemed to me unnecessary, but I did not interfere in a colleague's decision. He was admitted onto the medical ward, as I recall, sometime after lunch. In the evening he spiked a temperature and the next morning he was toxic and collapsed on the ward in a coma, which turned out to be due to septaecemia, with gross lung and kidney involvement. He was immediately resuscitated and intravenous antibiotics started. The microscopic examination of aspirated sputum showed gram +ve cocci and cultures including blood cultures grew Staph Aureus. A CT and MRI of his knee were carried out which showed a lesion in the distal femur that looked like an abscess.

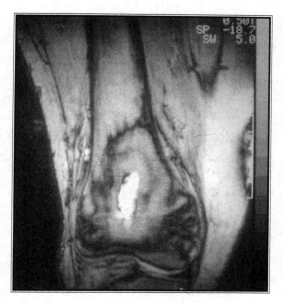

MRI of this patient's distal femur showing intraosseous abscess

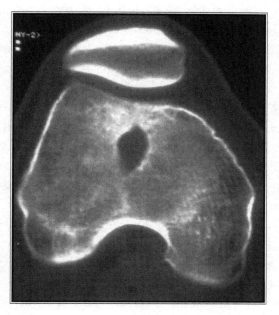

Transverse section his CT examination showing
intraosseous lesion (abscess)

They called me to do a needle aspiration of the bone which poured pus. I took him to theatre and carried out a formal drainage. He made an uneventful recovery. What disturbed me was that I was going to send him back to his base in the middle of nowhere and even in retrospect this seemed the most reasonable course, so I asked Dr. Freed, "Why did you admit him? What did you find that I didn't?" "Nothing," he replied. "But I know these guys. They are all highly positive "macho idiots" and telling them to take rest in camp is like talking to a wall. The only reason I admitted him was to ensure he rested up for 48 hours and took his medication."

And sometimes you have just got to examine the patient carefully!

Your two year old nephew, Ethan developed a temperature, which at times reached 40'C. In spite of antipyretics his temperature remained high, his Haemoglobin was low, his white cell count was high but within normal limits. After five days his parents wanted him seen by a Paediatric physician and we arrange for one of my colleagues, a Pediatric infectious disease specialist, Dr. Erez to see him. At this point he was apathetic, had an occasional cough, was not eating and occasionally complained of lower abdominal pains. He was to go to the Children's emergency room and be seen by ER doctor, Dr. Munro, first. She checked him and found both ears were inflamed. She took blood for cultures, full blood count and serology, and we were to wait for the results and for Dr. Erez to see him. At the last minute she decided for the sake of completeness to send him for a chest radiograph. This showed severe right upper and middle lobe pneumonia. On returning to clinical examination the signs were there. Dr. Erez put my fingers on his chest wall and the reverberations of his cries were indeed markedly increased on the right. Even I who hadn't examined a baby's chest for many, many years could detect this sign. After Ethan had recovered I rang Dr. Munro to thank her and let her know the follow-up on his progress. I asked her why, when she had found a cause (ear infection) for his temperature, did she sent him for chest x-rays? I think she was a little embarrassed she had missed the

So You're a Doctor Now

diagnosis because she rather hummed and hawed a bit and then said it protocol to do a chest x-ray in a child with a white cell count of more than 15,000 (Ethan's was 14,000) and as she was going to send him for an x-ray anyway she hadn't spent time examining his chest.

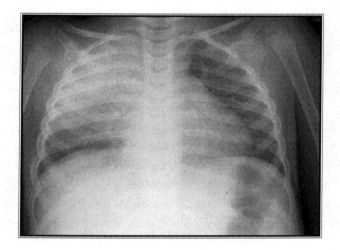

Ethan's chest radiograph -AP

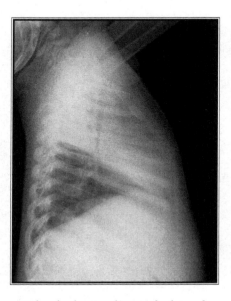

Ethan's chest radiograph- lateral

So, I still don't really know why she sent him for an x-ray. But it does raise a good point. Dr. Munro who had been working in the children's emergency room for more than 20 years, failed to pick up these signs. If she could miss them, it shows how easy it is when you find a cause of the temperature or whatever, to become complacent during the rest of the examination.

When should you order investigations? Obviously blood tests, radiograph, Computerized tomography, Nuclear magnetic resonance studies etc. cannot be ordered on every patient every time he attends clinic. And there is a downside of all these "unnecessary" investigations; they are expensive and a waste of precious resources, some are painful and unpleasant for the patient or even dangerous (e.g. Edward and the gastroscopy). Even a plain radiogram involves ionizing radiation. So when should you get, for example, a chest x-ray? When you think the patient has pneumonia or when you think he hasn't and you want to be sure? Unfortunately, I can't help you with this. I think it is partially the physician's personality. Some have confidence in their diagnoses and don't need so many investigations, some feel insecure and worried about missing something, and sometimes you know the investigation will be normal but the patient "needs" it as a psychological prop.

THE MIND AND PHYSICAL ILLNESS

I don't know which branch of medicine you will choose, but today Medicine is getting more and more specialized (jokes abound about right big toe and left big toe specialists). I first came across this when I was starting orthopaedics in Cardiff. In those days, most British orthopaedic surgeons treated most of the conditions, with perhaps an interest in hands, or joint replacement, or back problems. We were visited by an American surgeon, who not only just treated knee problems but only soft tissue problems of the knee. I couldn't believe it, but as time has passed this trend has continued. (Even I now only treat children's orthopaedic problems). Unfortunately many conditions cross the lines of these specialties. Many, for example, diabetes, hypertension, infections, degenerative diseases are multifactorial and many people are now so specialized they don't have the training to diagnose or treat anything outside their closed field. On the other hand they probably treat their chosen conditions far better than the old generalist ever could. Even in the old days, this problem occurred.

I was a junior registrar and we were doing the weekly consultant ward round, when we came to the bed of a middle aged lady. As we approached her bed she started to cry. The consultant asked her why she was crying. "Oh, doctor," she started. "I'm so depressed. I --." At this point he stopped her.

"Look, I have thirty orthopaedic patients to look after. I don't have the time, patience or desire to hear all your problems. You can be depressed at home, but here your problems are only orthopaedic. Now tell me what they are."

She sniffed a little, stopped crying and in a perfectly normal manner proceeded to tell us her symptoms. I have no recollection of what her orthopaedic problem was, but I do remember at the time, being impressed and agreeing with the consultant (I was near the beginning of my medical career). Certainly specialists are not and cannot be expected to be holistic therapists. Particularly surgeons, who have a very heavy patient load and responsibility simply do not have the time, inclination or learning to treat psychological or mental or even medical problems. On the other hand he could perhaps have expressed this a little more kindly with an instruction to arrange for her to see a psychiatrist on discharge from hospital.

It may be important to recognize the connection between the psyche and physical. This can often be very difficult.

Sometime ago, a traumatologist referred a fourteen year old boy, David, to me. He had sprained his ankle one month earlier and been treated in plaster for three weeks. However, one week after the plaster was removed he still was on crutches, complained of generalized pain around his injured foot and was unable to weight bear. At this time there was not much to find clinically and I gave him some good pain killers and told his mother to bring him back in one week (as I have mentioned, problems that don't progress along a normal time-line, turn on a "Red Light". A minor sprain after 3 weeks in plaster and 1 week to get back to normal should have been more than sufficient for the patient to be well on the way to recovery). The following week he came back, if anything, worse and no sooner had I touched the skin he started "owing and ooing". There is an uncommon condition (I had only seen it twice before) called autonomic sympathetic reflex dystrophy. After minor injuries patients develop severe pain, gross hypersensitivity, the skin becomes purplish and they are totally unable to use the limb. This condition is known in many cases to have a psychological connection. It is a horrible condition, difficult to treat and in the past patients have had the limb amputated and even committed suicide because of the severe continuous pain. I sent him to one of two centres that specialize in this condition. Over the next few months, I saw him and his mother frequently and got to know them well. His mother assured me there were no problems at home or school and indeed he did seem a well-balanced and a really nice kid. In spite intensive physiotherapy, drug and supportive psychotherapy he just got worse and worse. Several months went by without improvement so I asked the pain clinic specialists here in Beer-Sheva to help. In the end they admitted him to the intensive care ward, in order to monitor the huge quantities of morphine and ketamine he was requiring. However even the pain care specialists were at a loss. They told me they had never seen anybody in such severe pain that had required such high doses of powerful drugs which didn't even control the pain. It was during this time that I had proof of the interaction between the psyche and physical. Even when he was asleep and his leg was lightly touched he groaned in pain and had a withdrawal reflex. The only time this didn't happen was when the dose of ketamine was so high he was effectively anaesthetized. This was genuine physical pain. Nothing was working. By chance, I happened to be visiting him in the Intensive care ward at the same time that they told him and his mother he was being transferred to the Children's Pain Centre in Schneider Hospital, Petach Tikvah. At the time I didn't think anything of it but he smiled (I think it must have stuck in my memory because I had never seen him smile before).

Two weeks later I saw him, shopping in the mall near my clinic with his mother, and walking around without crutches completely normally. You could, as the saying goes, have knocked me over with a feather. They told me he had had treatment and psychological help and now he was cured.

I immediately went to the clinic and rang up Schneider. What was their secret? They felt it was the psychological help they had given him. They also told me when he was six his brother had developed leukemia and had been treated in the Children's Cancer Center in Schneider hospital (which is also the country's tertiary referral centre for Children's Oncology). His parents lived in Beer-Sheva and due to the long distance and time they had to spend with their ill son, David was sent to live with his grandparents. I don't know whether the Schneider psychologists had any special treatment that the previous centre didn't have. Maybe they did, but I do remember his smile when he heard he was being transferred to Schneider.

"That the wheel may turn and still be forever still"

"That the wheel may turn and still be forever still" is a quote from Murder in the Cathedral, a play by T.S.Eliot; Time goes by but nothing changes. There is something reassuring in knowing the safety and awareness of yesterday will be there tomorrow and yet, in a way, it is rather depressing.

After I discovered that the Universe, outside orthopaedics, is an interesting place, I took a course, along with 1st year students, in Geology at Ben Gurion University. Included in this course were day field trips to see different geological formations and a long weekend trip down to Eilat where within a small area there are many different rock strata. At the end of a wonderful inspiring day's geology, I was on the balcony of our accommodation after supper, drinking a glass of beer, sitting with my fellow students, and basking in nostalgia. I was back in time to my own student days, when many a happy hour was passed in similar repose. During the conversation, one of the students took out a packet of cigarette papers and proceeded to "roll-his-own". Once more, this took me back to my student days and my friend Colin. Colin had been a merchant seaman before starting medicine. He had been in charge of loading and discharging cargoes. As different types of cargoes are taken on and off at different ports, knowing which to put where so they didn't interfere with each other, yet take up the minimum of room and were layered to cause the minimum of disturbance and discharge/reloading at each port was a highly skilled job. However when containers became commonplace this skill became less necessary and Colin found the job boring. He brought with him the habit of rolling his own; take out a 'Rizla' paper; sprinkle on tobacco, back and forth roll between thumbs, index and middle fingers, lick to right, lick to left on adhesive strip, final roll between fingers. I asked him once why he rolled his own instead of buying readymade ones. Partly, he told me, out of habit but mostly because it was much cheaper. None of us students had much money and this made perfect sense. Here, thirty odd years later were students in the same position, saving money rolling their own. Oh, yes, nostalgia was good. After taking a few puffs, he passed the cigarette to the student next to him. Wow, I thought, we had it lucky, we were poor, but we didn't have to share cigarettes. After a few puffs the second student passed the cigarette to his neighbour. Something, I thought, is decidedly odd. At this point the first student started giggling uncontrollably at a rather inane and not particularly amusing comment. The penny dropped. The wheel may turn and sometimes it does move a bit.

A "Rollie" showing the rolling action between thumb, index and middle fingers.

Pride

The following story really has nothing to do with medicine, but I put it in because it's a nice story and great moral.

It was February 2005, the Omer Run.

Omer run

This is a day's sporting event whose highlight is a ten kilometre run around the village. I was taking part in this run, but as soon as we had started I felt I wasn't going to be able to finish. Anyway I did manage somehow to continue. The field gradually settled out; the front runners, the middle pack and the tail enders with whom I was running. I was running with a thin wiry guy in a grey running shorts and vest. I say I was running with him, but actually he was running faster than me, except when we arrived at a hill, or rather mild incline as there are no actual hills here in Omer. Then he stopped running and walked and I passed him. Once on the flat again, he passed me. And so we ran the whole race, neck and neck. As it happened about five hundred metres from the end was a long slope and I just managed to overtake him as we reached the top. With two hundred metres left, I was determined to beat him and I ran as though the devil himself was behind me. Not only did I beat him, I achieved my best time ever (fifty seven minutes), and I was extremely pleased with myself. Afterwards I went up to him and during the conversation I said, "You are a better runner than I am, and I don't understand why you stopped and walked up the inclines."

"Oh," he said, "Where I come from it's all flat and I just didn't have the breath to run up the slopes." That seemed reasonable and we continued talking. During the conversation he asked me "How old are you?" "Oh, I'm 54 already." I answered proudly. "How old are you?"

"74."

Pride certainly has a way of kicking a man when he's up!

RESEARCH

The ABC Travelling Fellows

I don't know if you will be drawn into the world of research. Some people are, some aren't. I hope you will be because the stimulus of looking for new facts, treatments, results or just new observations is just that- a terrific stimulus.

If you are ensnared by the lure of research and you discover something new that will change our treatments be prepared for a long uphill battle, possibly vilification and a lot of plain ignorant nastiness. People have great difficulty admitting they have been doing things wrong half their life time.

I told you about Cardiff and the development Carbon fibre plates and the difficulty we had in getting them accepted. They had some technical problems and these were jumped on to kill the project.

However, one year Cardiff was chosen to be visited by the ABC Travelling Fellows. This extremely prodigious fellowship is awarded only to a few very top, brightest orthopaedic surgeons from America, Britain, Canada, New Zealand and South Africa who are under 45 years old, and have proven themselves but are still near the beginning of their careers. In odd years the American and Canadian group tours centres in Australia, New Zealand, South Africa and United Kingdom and in the even years the reverse takes place.

The groups tour six or so centres and it is a great honour and privilege to be chosen to receive them.

The major research theme in Cardiff was bone healing and the development of Carbon fibre plates which were designed to "use" biology to help fracture healing rather than the rigid fixation of the AO philosophy which inhibits callus (new bone). We were anti AO (this is the system of fracture fixation I mentioned in the section "Carbon fibre plates").

The Senior lecturer, Mr. Thorton who was running the research was a wonderful energetic speaker and he presented the Cardiff fracture philosophy, the carbon plate, tearing the AO concept to threads, to the Travelling Fellows. I sat there, listening, thoroughly enjoying the lecture and extremely proud of my orthopaedic Centre. I had no doubt these top American and Canadian surgeons were getting a "Road to Damascus" experience and I was sure Mr. Thornton would get a standing ovation. I was wrong. One of the surgeons arose and said "Who the hell do you think you are that you criticize the AO like that?!" I was flabbergasted not only by the vehemence of his language, not only by his refusal to see the facts put in front of him, but by the fact that any criticism of his own belief (I mentioned before the AO system had swept the orthopaedic world in the 1950's and 60's) was inadmissible. And these were the top guys sent to see and learn how other centres did things!

Prepare your project

As a Registrar in Cardiff, I remember going to Professor McKay at least once a week or so with some new idea for a project I wanted to do. He tended to reject these one after the other and after a while I was getting discouraged, so I asked him why he kept rejecting my ideas. "Charles," he said. "Ideas are ten a penny. What you need to do is first think about them, look up the literature, see whether it's been done before and if so maybe you'll have a different slant or approach, see whether the idea is practical; do you have access to the equipment? Do you have the specialized or technical knowledge to carry out the project, is it feasible in the available time, finances, etc. When you've done all that, come to me again and we will discuss it."

Since then if I go to meetings with a project I want to present, I first make a PowerPoint presentation and try to learn up as much as I can about the Pros and Cons before discussing it. Nevertheless, in meetings of all sorts you can guarantee, no matter how well prepared the presenter is, someone will volunteer some idea he had thirty seconds previously. Although it is possible, he could instantly think of something that over the previous months all your thinking, searching, reading and preparing had missed, it is highly unlikely. I mention it because it happens frequently, is annoying and a little insulting, but just smile, be polite and explain why it is or isn't a possibility.

Here are a couple of examples. I wanted to look at the effect of heat on bone healing:- the concept being that if one warmed up the fractured limb, there would be increased blood flow and quicker or more solid union. After four months looking into this, arranging a rabbit experiment including all the specialized heating and assessment equipment, I presented the idea at our monthly research meeting. Thirty seconds after I had made my presentation, one of the doctors piped up. "Yes, and you should test the effect of cold to see if it reduces healing."

I was at a meeting of the Environment Committee of our local council and we were discussing the building of a cycle route. One of the members was really annoyed. "They are installing new underground water and electricity infrastructure along the roads and pavements, so why didn't anybody speak to the contractors and get them to make a cycle path at the same time. It would have hardly added to the expense."

He hadn't prepared a route for the bicycles, he hadn't looked up the legal requirements, he hadn't looked up the protocols of what sort of cycle paths were suitable, effect on traffic, widths, protection for cyclists from passing cars, effect on home owner's drives, car parking, materials, what markings are required, etc. Nothing. But he did waste the whole of the rest of the meeting pontificating about it. Don't be one of these people!

Different people interpret the same thing differently.

You have to remember that people have different education and learning and will not see the same situation as you see it. The old adage about the optimist seeing a glass half full while the pessimist sees it half empty illustrates this well, and because you will come across this so often, it is important to try as much as possible to use words and explanations with regard to how the other person will interpret them. Don't ask a patient if he has angina pectoris, rather does he suffer from chest pain on exercise.

The following story illustrates this well.

Stepan Petriev trained as a Paediatric surgeon but now heads a startup company to develop a number of his inventions. One of these is a stapler to anastomose the intestines after resection. In the "old days" this was done by a special technique involving inverting the bowel ends and suturing them in three layers; a time consuming and difficult process. It is essential there are no leaks, but at the same time no narrowing of the lumen. At the time there were two companies producing staple guns. Petriev invented a much improved version. He received a grant from the Ministry of Science to develop it. This grant covered the costs of manufacturing a prototype and the salary for six months of an engineer. At the end of this time it was expected he should have a working prototype to commercialize. After two months he asked his engineer if it was ready yet.

"Oh, no we are having a lot of difficulty with some of the parts."

A further two months past and again he was informed by his engineer that he had not been able to overcome the difficulty. Petriev asked him what exactly this difficulty was.

"Well, cutting the metal parts isn't a problem but plastic is very soft and trying to cut it to the tolerances, 0.01mm, as required by the plans, is extremely difficult."

"Why do you have to cut the plastic to such accuracy? 0.01mm is almost microscopic."

"Because that is what is written on the plans."

Petriev who wasn't an engineer had been told that moving metal parts, being hard, need to be very accurately machined to a tolerance of 0.01mm. He had assumed that plastic, being much softer, would if anything, be easier to machine, so he had written down the same tolerance for these parts when preparing the blueprints.

"Look," he said to the engineer. "It doesn't matter if these particular parts are +/- 0.5cm!. Just cut them and let's get on making the prototype!"

Anyway, after this initial delay, he did manage to produce a prototype just before his six month grant ran out.

He took it to one of the two companies (I forget which one) who marketed intestinal staple guns. They were very impressed by it and freely admitted it was better than theirs.

"So you will buy it from me?" asked Petriev.

"No," they replied

"Why not?" asked Petriev, somewhat surprised.

"Well, it's like this. We have about 50% of the market and our competitor also has about 50%. If we take yours, we will have to file patents, which is very expensive. We will have to retool our factory and retrain our workers, which is very expensive. We will have to apply for FDA and EC approval which is very time consuming and expensive. We will need to do clinical trials to prove it is safe and that it is indeed better, which is very difficult and very expensive. We will then have to give it to our agents who will have to persuade surgeons to give up their present system with which they are comfortable and go to the expense of re-equipping with this new stapler. Now assuming we did all this and were successful in all stages without any hiccups, we could probably increase our share of the market to 70% or 75%, but unfortunately this simply doesn't justify the risk."

Petriev saw the stapler as a wonderful advance in the treatment of surgical patients.

The engineer saw it as a difficult technical problem of material machining.

The surgical instrument company saw it in terms of financial profits.

Making an original discovery is an exciting an experience you can ever hope to achieve. The dictionary defines "original" in two ways; "First or earliest", and "not dependent on other people's ideas, inventive or novel". It is the second of the two definitions I think is the important one. It really doesn't matter if one, two or a hundred people have made the observation before you if you find it independently.

Digital artery aneurysm

When I was a Senior Registrar, I operated on a woman who presented with a painful swelling in one of her fingers. It turned out to be an aneurysm arising from a digital artery. I tried to read up about this condition but found nothing in any books, or in the literature. The pathologist had never seen one and we decided to write a case report about the condition. A deep search of the literature again revealed no further information. In those days literature search meant looking up keywords in the volumes of Cumulated Index Medicus, volume by volume year by year; a very time consuming task. We found nothing until we were about to send off the article when the latest set of the Index arrived and we found a similar case had been reported two months earlier. Was our discovery worth less? We did not get the credit for being the first, but it still was an original observation.

Meteorites

Let me recount the story of Widmanstätten and Thompson. As you know I have recently taken an interest in iron meteorites. They are intrinsically beautiful and the physics and chemistry surrounding their formation and life cycle can truly and figuratively speaking be described as "out of this world". As they cooled over millions of years from about 900 to 400 'C, the alloys of iron and nickel (kamacite and taenite) slowly crystallized out into separate areas, which can be seen when the meteorite is oxidized by heating or by chemical etching, and they are known as Widmanstätten patterns. Count Alois von Beckh Widmanstätten had an interest in science but was not a professional scientist. At the time he was the director of the Imperial Porcelain Company in Vienna. How he came to possess an iron meteorite and why he was heating it is unknown, but he did have such a meteorite and he did heat it. He noted the different colours produced by the oxides of kamacite and taenite, which produced beautiful patterns on the meteorite. He was not a scientist, but discussed his finding with his friends. One of these, Schreibers, was particularly interested in meteorites and in 1808 Schreibers published the findings, giving credit to Widmanstätten. Some years earlier, however, an English geologist, G Thompson had made similar observations while working in Naples. Unfortunately at that time, Southern Italy was in the crisis of civil war and political unrest. The messenger with whom he sent his findings was murdered and his letters lost. He did, however, manage to publish his work in French in a small little known Belgium journal, Bibliothèque Britannique, in 1804. When Napoleon entered Naples in 1806, Thompson fled to Sicily where he died in the same year at the age of 46. The meteorite, shown in the figure below called Campo del Cielo, fell in what is now Argentina. These meteorite fragments were known and used by the Inca civilization and then in 1576, by the Spaniards. Although there is no record, I am sure an Inca or Spanish blacksmith (and many of them) also noted these patterns independently.

Campo del Cielo meteorite, found Santiago Estero, Argentina.

Etched slab of campo del Cielo iron meteorite, found Santiago Estero, Argentina, showing Widmanstätten patterns of Kamacite and Taenite.

Family

The biggest price we as doctors have to pay is the loss of time with our families. This is particularly true at the beginning of your career when you are learning, working long hours of on-duty, covering days and nights. This is just when you have a young family and when they need you most. I will never forget talking to a retired general surgeon in Swansea. I was a young and very enthusiastic surgeon at the time and loved every exciting, interesting minute of my time in the hospital. I felt awfully sorry for this man who had now severed his links with the profession. I asked him how he could possibly cope with this (I assumed it must be like withdrawing heroin from an addict). His reply was like a thunderbolt (which, perhaps, is why I still remember it so clearly). "Retirement," he said, "is wonderful. I can now do with my grandchildren all the things I missed out on with my own children." He looked away sadly, sighed and said half to himself, half to me. "Oh, how I regret that lost time." Was I heading on the same path? Would I look back on my life 40 years on and regret how I spent it? I know the pressures/ hours of work prevented me from spending as much time with you as I would have liked; life is a compromise. Yes, I made mistakes, not on purpose, but due to lack of experience together with trying to build a career. Unfortunately, nobody knows how to be a parent until you suddenly are one. I remember my mother telling me the following tale.

It concerned a young man, who was constantly in trouble with the police for thefts from shops, and burglaries. He was an only child and was the apple of the eye of two doting parents. They both had very demanding jobs but were, in consequence, well off; most of this wealth being directed to the happiness, education and well-being of their son. During his childhood he wanted for nothing, but in adolescence and early youth turned to theft.

In desperation the parents went, with their son, to see a family psychiatrist. The psychiatrist spoke first to the parents. They were patently very worried and distressed about their son and volunteered all the information they could. They were asked about their son's childhood and how they had brought him up. Whatever he wanted, they were only too happy to give him. They had got him the best nannies, sent him to the best and most expensive schools and tried to set a good moral example. As I said before they doted on him, gave him everything he wanted and had ever asked for; toys, sports sessions, hobbies, technological gadgets etc. The psychiatrist then spoke to the son. In the course of the interview, he said to the doctor, "of course my parents never really loved me." The psychiatrist, somewhat surprised to hear this, asked him to explain. "Well," replied the son, "one of my earliest memories was playing out in the street with all my friends. When six O'clock came, all their parents came out to call them in for supper. But my parents never did; they just left me out there, in the street. They couldn't care less what happened to me on my own." When challenged with this, his parents answered thus. "Of course we didn't call him in. He was enjoying himself so much playing. Whenever he wanted to come in for supper, that was fine with us. We weren't like those other parents, spoiling their children's fun just because it was convenient for them to eat supper at six O'clock."

Recently the following story came to my attention. There lived in our town a very well-known and highly regarded historian and author of several books and many scholarly articles. He had one child, with whom he did not get on. The son left home to live in America, never visited and rarely phoned his parents. They had no other family here. A few months ago, the father became terminally ill (at the age of ninety two) and within a few weeks he died. His son did not come to visit him during his last days, although he did come to the funeral. At his stone setting, he told some stories of how his father used to write his articles and books, and some analogy with the paintings of Cezanne. He was very respectful of his father's academic achievements and totally devoid of any warmth. What could be sadder? I was heartbroken for his father; in spite of holding the respect of the academic community, he had nothing. I can imagine no greater pain or distress than when the baby you held when first born, the child with whom you played and to whom you read bed time stories, who you saw learning to walk and talk, who you watched grow and develop, as an adult wants nothing to do with you.

Against this, is the story of Ludmila, who is now 67, immigrated to Israel from Russia some twenty odd years ago. She was a teacher in Russia, but she was unable to find employment in her profession. So now she gets up at half past five in the morning, five days a week, takes a three quarter of an hour bus ride to Omer to work as a cleaner in other peoples' houses. Her husband, by second marriage, has been unable to work for several years as he suffers from severe Parkinson's.

So in addition to cleaning houses, she has to look after him. Her only daughter, by her first marriage, son-in-law and grandchildren still live in Russia, where I gather they are reasonably comfortably off. The daughter works with computers, the son-in-law is a computer typesetter for a local newspaper and the granddaughter is finishing University. Ludmila saves money from the small amount she earns to send them. She also sends various packets of clothes and food whenever she can. Once every few years she visits them or sends them a ticket to visit her here. They talk daily on Skype and in spite of the distance, they are very close.

Society may admire the historian, but I admire Ludmila.

Unfortunately, nobody has innate knowledge of how to be a parent until suddenly you are one. Yes, you make mistakes, not on purpose, but due to lack of experience coinciding with trying to build a career. When I look back now I would do things differently, but I can't. So please forgive all my mistakes and put them down to inexperience and, because the bottom line is we ***do*** love you with a passion you will only understand when you have children of your own.

Where children are concerned, I think there is a contract with the Creator. I know you don't agree with what I am about to tell and it is totally contrary to Eastern and many other philosophies, but this is my view and I'm sticking to it!

He gives us children and in exchange it is our duty, our obligation and responsibility to bring them up and do everything that we consider is the best for them. Now, we don't have to get it right, but we do have to do what we *think* is the best. If we think it is best to be strict and stern with our children, we have no choice, so must we act. If we think it is best for them not to watch television, not have a mobile phone, never eat junk food, be in bed by seven o'clock then that is what we *must* do. If giving them almost everything they ask for, allowing them to stay up as late as they like etc. is best for them in our opinion, then this is the regime we <u>must</u> follow. Obviously how we bring up our children will vary from parent to parent and from child to child. For us these contractual obligations turned out to be the most satisfying and pleasurable of any activities in our lives. In return what does the child owe the parent? Absolutely nothing. They are under no obligation to their parents whatsoever, not to follow in their culture, profession, religion, ethics or even say "thank you", and certainly not *obligated* to look after us in our old age. Of course, it is nice if they do, and one hopes they will love and have affectionate and grateful feelings etc., but they don't *have* to. But, as they say, there is no such thing as a free lunch. So what is their side of the contract? It is this; I will do everything in my power to do the best for you, and in return you *must* do the best in your power for my grandchildren.

IT'S NOT INTERESTING

There is no such thing as "it's not interesting".

One of the problems with medicine is that it is so interesting and absorbing, it is very easy to become totally engrossed by it to the exclusion of all else. This happens to many doctors (put two in a room and you can almost guarantee they will talk medicine. In a way it's great that this does happens. It is only by keeping a love and passion that a doctor can retain his commitment to his patients). When I was a senior registrar in Oxford, we met and became friendly with a historian from Haifa. He was in Oxford to research a book, I forget the exact title but it was something like, "The influence of Islam on Jewish Culture in Persia, 610AD to 925AD". I had spent my medical student years with other medical students and my time training as an orthopaedic surgeon with other medical staff and it was not until the year in Oxford was I exposed socially to people other than medics. Nevertheless I was totally flabbergasted that anybody would want to read even a short article on this subject and was certain nobody would want to buy a book about it. So how could anybody spend a year of his life, move his family to a foreign country in order to write such a book? It took me a long time pondering about this and it was only when I was in a book shop, thinking about it, that I saw the plethora of books on so many varied and different subjects that I came to the inevitable conclusion that there is <u>nothing</u> that isn't interesting. Unfortunately life is so short we don't have time to go deeply into all the other subjects and extract the interest lying deeply hidden within them. But it did and does bother me; the world is so full and we are missing so much.

Advice No 1

Yes, medicine should grasp most of your attention and interest, but keep your eyes open, look for other interesting aspects of your world and the Universe.

Advice No 2

Part of my present to you is a picture of bleeding therapy that used to be standard treatment for thousands of years. It has always made me very angry because even if the apprentice doctor had been taught rigidly that this was the correct treatment, after carrying it out once, or twice or three times, he surely couldn't help noticing that it didn't work, and in some cases the patients got worse, even died. I have no doubt in many cases, bleeding was enough to push the patient over the brink and must have been lethal to many patients. George Washington developed either pneumonia or quinsy and had difficulty breathing. His doctors bled him over a quart of blood and by their own admission this is what caused, or greatly contributed to his death. And yet these doctors continued to bleed their patients. Thousands of their predecessors and contemporaries bled patients throughout their careers and passed it on to the next generation, who in turn continued to pass it on.

So You're a Doctor Now

Barber surgeon bleeding a patient attrib. JJ Horemanns
(author's copy)

It wasn't that long ago (just over one hundred and fifty years) that it finally went out of usage. Surely *someone* must have noticed it didn't work and was dangerous! And what about all the other nonsense quack treatments they used to use, such as cupping. So advice No 2, learn the art of REAL medicine (history examination, relevant investigations and effective treatments).

Keep an open mind about treatments. Are we using any such treatments unaware they have no scientific backing and if we looked properly they would have no more effect than a placebo? If a patient reacts badly or not at all to a treatment is it because of an individual poor response or is the treatment not really effective?

And why did the population put up with it? Why did they continue to go to see these doctors? I think the answer lies partly in the previous stories. People want care and attention, and yes of course they want to get better, but just the attention and having some treatment, quack or genuine that *they* believe in, is at the very least an important part of the solution. What people want, expect and indeed are entitled to is the best possible treatment you can give them. That is going to vary depending where you are working and what resources you have. In the 17th century NOT bleeding a patient would have been negligent because it was considered the best available treatment. Similarly you can only treat a patient with the best resources at your disposal, which will vary depending where you are working. In small peripheral villages the doctors don't have easy access to radiological or bacteriological laboratory facilities so they are justified, for example, in giving antibiotics for <u>suspected</u> infections. If you work in the accident and emergency department with easy access to laboratory and radiological facilities, giving antibiotics, just in case, is much less justifiable.

Advice No 3

Always, always show the patients you care about them, and you are happy to see them. Even when you are tired, fed up, irritable, nothing seems to be going right and you have all sorts of personal problems, smile and be glad to see each patient and stop them worrying, let them know they ARE going to get better. This is why the "Do not disturb" notices on the nurses' uniforms is so disturbing. The philosophy behind it is so alien and the exact opposite of what medicine is about.

Advice No 4

I came across this from an Internet blog.
"There is only one reason children estrange themselves from their parents. They do not feel loved." I think this is absolutely correct.
They may, in fact, be deeply loved, but they *feel* they aren't. I think children need and crave their parents' attention.
Do as much as possible with your kids, from as young an age as possible. When you are washing up, making salad, cutting bread, cleaning the car, doing the garden, taking pictures, let them do it with you. Will they drop and break the camera? Yes, possibly, but it is very unlikely; take the risk. Will you miss out on capturing the glorious

Kingfisher diving into a clear stream at sunset? Yes; so what! You come home tired after work; you want to eat supper, rest a bit, and watch the news or the World cup on TV. Is that more interesting than reading about how 'Duffy Driver overslept and kept everybody waiting at the station for the little red train', for the twentieth time? Yes, of course it is; but read about Duffy Driver. One of my earliest memories was of my fifth birthday party. It was the days before television. Suddenly my dad produced a cine projector and showed some cartoons to all the children. It was something nobody else had ever done. All the children were extremely excited and it made the party an enormous success. He didn't say anything beforehand; he just went out and somehow obtained the unobtainable. He did it for us because he cared for us and it will always remain one of the highlights of my life.

On one occasion when I came back from England, I bought you a beautiful model of a sailing ship to make. It was large and very expensive (it was £80 in those days). But I didn't make the time to build it with you. I should have done and like the general surgeon, I regret it. So when you have kids, give them as much of your time as you possibly can.

Well. I think I have more or less finished. I am near retiring now, (well in a few years' time). You are just starting out. What have I learnt? I think it is best summed up by a final story.

The Chair

I was talking to a fairly young paediatrician who had trained in Soroka. As you know I spent the first five years when we returned to Israel working in Soroka and developed a very close relationship with the Paediatric department (being a paediatric orthopaedic surgeon) so naturally the conversation drifted to the various doctors with whom we had both worked. We started to discuss Dr. Erez (who had diagnosed Ethan's pneumonia). Dr. Erez had built up quite an "empire" in the infectious diseases world and made Soroka a world center for many clinical trials for various vaccines, such as pneumococcus and haemophilus. He has written many papers, spends more time abroad giving lectures, sitting on committees and chairing several Infectious diseases societies than he does in the hospital. Anyway, during the time he actually was in the hospital he was taking the students on a teaching round (the doctor with whom I was conversing had been one of them at the time). They entered one of the rooms, which, like all the paediatric wards, was overflowing with children and their families. As they entered, he said, "One moment," and walked out to return one minute later with a chair for the child's mother. Once she was seated, he launched into the lesson regarding the child's illness. The doctor with whom I was talking said she couldn't remember any details of the illnesses she was taught about on that day, but the incident with the chair had remained outstanding in her memory. Why I particularly like this story, is because it highlights what medicine is actually all about. It should be a subconscious reflex, a caring reflex, a desire to help people be more comfortable, free of pain, deformities, disabilities and ill health. Dr. Erez's task was to teach the students about the illnesses that were found on the ward that day, but he simply was unable to start until he had made the mother as comfortable as possible.